TWENTY FIRST CENTURY
science

Project Directors

Angela Hall   Emma Palmer

Robin Millar   Mary Whitehouse

Editor

Mary Whitehouse

Authors

Robin Millar   Elizabeth Swinbank

David Sang   Carol Tear

THE UNIVERSITY of York

THE SALTERS' INSTITUTE

Nuffield Foundation

OCR
RECOGNISING ACHIEVEMENT

OXFORD
UNIVERSITY PRESS

Official Publisher Partnership

Contents

How to use this book

Welcome to Twenty First Century Science. This book has been specially written by a partnership between OCR, The University of York Science Education Group, The Nuffield Foundation Curriculum Programme, and Oxford University Press.

On these two pages you can see the types of page you will find in this book, and the features on them. Everything in the book is designed to provide you with the support you need to help you prepare for your examinations and achieve your best.

Module Openers

Why study?: This explains why what you are about to learn is useful to scientists.

Find out about: Every module starts with a short list of the things you'll be covering.

Ideas about Science: Here you can read about the key ideas about science covered in this module.

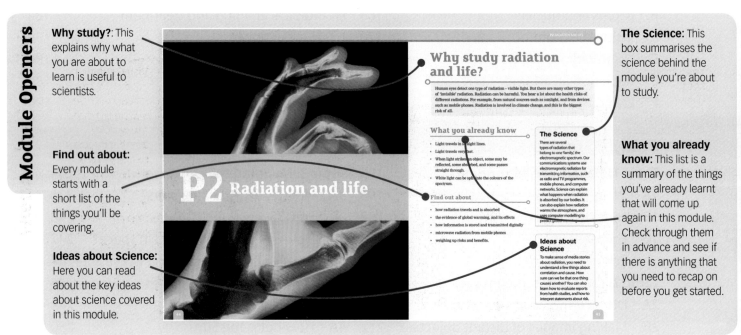

The Science: This box summarises the science behind the module you're about to study.

What you already know: This list is a summary of the things you've already learnt that will come up again in this module. Check through them in advance and see if there is anything that you need to recap on before you get started.

Main Pages

Find out about: For every part of the book you can see a list of the key points explored in that section.

Worked examples: These help you understand how to use an equation or to work through a calculation. You can check back whenever you use the calculation in your work to make sure you understand.

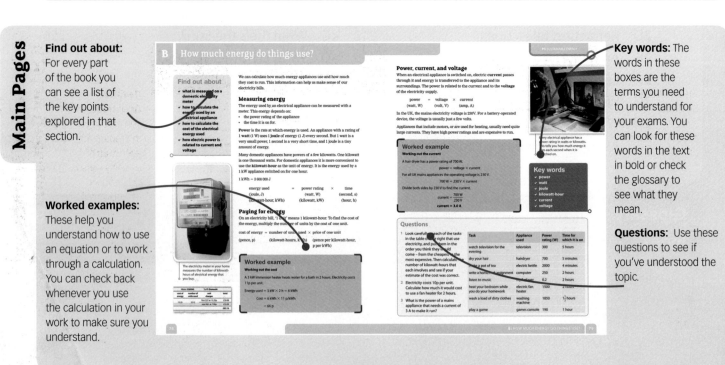

Key words: The words in these boxes are the terms you need to understand for your exams. You can look for these words in the text in bold or check the glossary to see what they mean.

Questions: Use these questions to see if you've understood the topic.

You should know: This is a summary of the main ideas in the unit. You can use it as a starting point for revision, to check that you know about the big ideas covered.

Visual summary: Another way to start revision is to use a visual summary, linking ideas together in groups so that you can see how one topic relates to another. You can use this page as a starting point for your own summary.

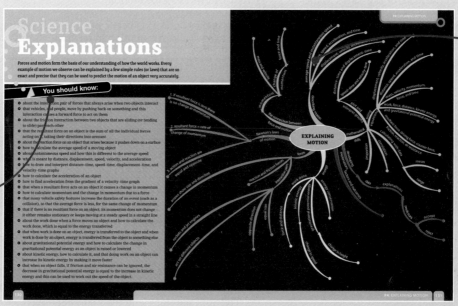

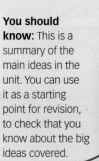

Ideas about Science: For every module this page summarises the ideas about science that you need to understand.

Review Questions: You can begin to prepare for your exams by using these questions to test how well you know the topics in this module.

Structure of assessment

Matching your course

What's in each module?

As you go through the book you should use the module opener pages to understand what you will be learning and why it is important. The table below gives an overview of which main topics each module includes.

P1
• What do we know about the place of the Earth in the Universe? • What do we know about the Earth and how it is changing?

P4
• How can we describe motion? • What are forces? • What is the connection between forces and motion? • How can we describe motion in terms of energy changes?

P2
• What types of electromagnetic radiation are there? What happens when radiation hits an object? • Which types of electromagnetic radiation harm living tissue and why? • What is the evidence for global warming, why might it be occurring, and how serious a threat is it? • How are electromagnetic waves used in communications?

P5
• Electric current – a flow of what? • What determines the size of the current in an electric circuit and the energy it transfers? • How do parallel and series circuits work? • How is mains electricity produced? How are voltages and currents induced? • How do electric motors work?

P3
• How much energy do we use? • How can electricity be generated? • Which energy sources should we choose?

P6
• Why are some materials radioactive? • How can radioactive materials be used and handled safely, including wastes?

P7

• Naked-eye astronomy
• Light, telescopes, and images
• Mapping the Universe
      • The Sun, the stars, and their surroundings
      • The astronomy community

How do the modules fit together?

The modules in this book have been written to match the specification for GCSE Physics. In the diagram below you can see that the modules can also be used to study parts of GCSE Science and GCSE Additional Science.

		GCSE Biology	GCSE Chemistry	GCSE Physics
GCSE Science		B1	C1	P1
		B2	C2	P2
		B3	C3	P3
GCSE Additional Science		B4	C4	P4
		B5	C5	P5
		B6	C6	P6
		B7	C7	P7

GCSE Physics assessment

The content in the modules of this book matches the modules of the specification.

The diagram below shows you which modules are included in each exam paper. It also shows you how much of your final mark you will be working towards in each paper.

	Unit	Modules Tested			Percentage	Type	Time	Marks Available
Route 1	A181	P1	P2	P3	25%	Written Exam	1 h	60
	A182	P4	P5	P6	25%	Written Exam	1 h	60
	A183	P7			25%	Written Exam	1 h	60
	A184	Controlled Assessment			25%		4.5–6 h	64

Command words

The list below explains some of the common words you will see used in exam questions.

Calculate
Work out a number. You can use your calculator to help you. You may need to use an equation. The question will say if your working must be shown. (Hint: don't confuse with 'Estimate' or 'Predict'.)

Compare
Write about the similarities and differences between two things.

Describe
Write a detailed answer that covers what happens, when it happens, and where it happens. Talk about facts and characteristics. (Hint: don't confuse with 'Explain'.)

Discuss
Write about the issues related to a topic. You may need to talk about the opposing sides of a debate, and you may need to show the difference between ideas, opinions, and facts.

Estimate
Suggest an approximate (rough) value, without performing a full calculation or an accurate measurement. Don't just guess – use your knowledge of science to suggest a realistic value. (Hint: don't confuse with 'Calculate' and 'Predict'.)

Explain
Write a detailed answer that covers how and why a thing happens. Talk about mechanisms and reasons. (Hint: don't confuse with 'Describe'.)

Evaluate
You will be given some facts, data, or other kind of information. Write about the data or facts and provide your own conclusion or opinion on them.

Justify
Give some evidence or write down an explanation to tell the examiner why you gave an answer.

Outline
Give only the key facts of the topic. You may need to set out the steps of a procedure or process – make sure you write down the steps in the correct order.

Predict
Look at some data and suggest a realistic value or outcome. You may use a calculation to help. Don't guess – look at trends in the data and use your knowledge of science. (Hint: don't confuse with 'Calculate' or 'Estimate'.)

Show
Write down the details, steps, or calculations needed to prove an answer that you have given.

Suggest
Think about what you've learnt and apply it to a new situation or context. Use what you have learnt to suggest sensible answers to the question.

Write down
Give a short answer, without a supporting argument.

Top Tips
Always read exam questions carefully, even if you recognise the word used. Look at the information in the question and the number of answer lines to see how much detail the examiner is looking for.

You can use bullet points or a diagram if it helps your answer.

If a number needs units you should include them, unless the units are already given on the answer line.

Making sense of graphs

Scientists use graphs and charts to present data clearly and to look for patterns in the data. You will need to plot graphs or draw charts to present data and then describe and explain what the data is showing. Examination questions may also give you a graph and ask you to describe and explain what a graph is telling you.

Reading the axes

Look at these two charts, which both provide data about daily energy use in several countries.

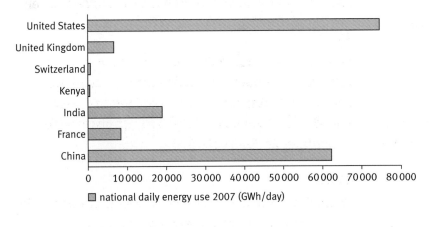

national daily energy use 2007 (GWh/day)

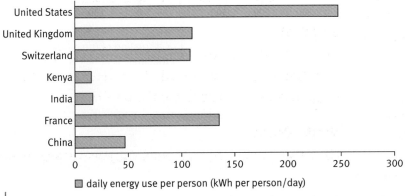

daily energy use per person (kWh per person/day)

Graphs to show energy use in a range of countries, total and per capita.

Why are the charts so different if they both represent information about energy use?

Look at the labels on the axes.

One shows the **energy use per person per day**, the other shows the **energy use per day by the whole country**.

For example, the first graph shows that China uses a similar amount of energy to the US. But the population of China is much greater – so the energy use per person is much less.

First rule of reading graphs: read the axes and check the units.

Describing the relationship between variables

The pattern of points plotted on a graph shows whether two **factors** are related.

Look at this graph.

There *is* a pattern in the data – quite a smooth curve; as the speed of the car increases, the stopping distance increases.

But it is not a straight line, so we can say more than that. The stopping distance increases more at higher speeds. The slope of the graph – the **gradient** – increases as the speed increases.

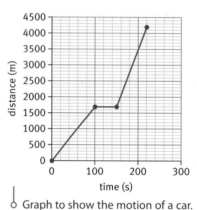

Graph to show the relationship between the stopping distance of a car and its speed.

Look at this graph, which describes the motion of a car:

How many different gradients can you see?

There are three phases to the graph, so you should describe each phase, including **data** if possible:

- The distance travelled **increases rapidly** for the first 100 seconds when it has reached 1700 m.
- For about the next 50 seconds its **distance does not change – it is stationary**.
- The distance then **increases rapidly** over the next 70 seconds – the line is steeper, which means the car is **moving more quickly**, from 1700 m to 4200 m.

Graph to show the motion of a car.

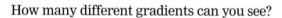

Calculation of the gradient of the line.

Calculating the gradient

The gradient of the distance–time graph gives the speed of the car.

$$\text{speed} = \frac{\text{distance travelled}}{\text{time taken}}$$

When the car is travelling fast it travels 1800 m in 50 seconds.

$$\text{speed} = \frac{1800 \text{ m}}{50 \text{s}} = 36 \text{ m/s}$$

Second rule of reading graphs: describe each phase of the graph, including ideas about the meaning of the **gradient** and other **data** including **units**.

Is there a correlation?

Sometimes we are interested in whether one thing changes when another does. If a change in one factor goes together with a change in something else, we say that the two things are **correlated**.

The two graphs on the right show how global temperatures have changed over time and how levels of carbon dioxide in the atmosphere have changed over time.

Is there a correlation between the two sets of data?

Look at the graphs – why is it difficult to decide if there is a correlation?

The two sets of data are over different periods of time, so although both graphs show a rise with time, it is difficult to see if there is a correlation.

It would be easier to identify a correlation if both sets of data were plotted for the same time period and placed one above the other, or on the same axes, like this:

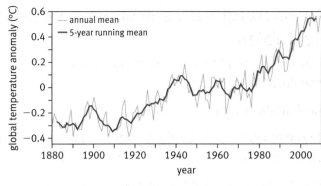

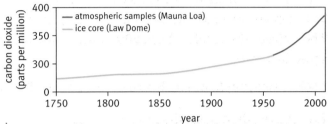

Graphs to show increasing global temperatures and carbon dioxide levels. Source: NASA.

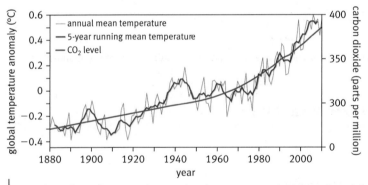

Graph to show the same data as the above two graphs, plotted on one set of axes.

When there are two sets of data on the same axes take care to look at which axis relates to which line.

Third rule for reading graphs: when looking for a correlation between two sets of data, read the axes carefully.

Explaining graphs

When a graph shows that there is a correlation between two sets of data, scientists try to find out if a change in one factor causes a change in the other. They use science ideas to look for an underlying mechanism to explain why two factors are related.

Controlled assessment

In GCSE Physics the controlled assessment counts for 25% of your total grade. Marks are given for a practical investigation.

Your school or college may give you the mark schemes for this.

This will help you understand how to get the most credit for your work.

Practical investigation (25%)

Investigations are carried out by scientists to try and find the answers to scientific questions. The skills you learn from this work will help prepare you to study any science course after GCSE.

To succeed with any investigation you will need to:
- choose a question to explore
- select equipment and use it appropriately and safely
- design ways of making accurate and reliable observations
- relate your investigation to work by other people investigating the same ideas.

Your investigation report will be based on the data you collect from your own experiments. You will also use information from other people's research. This is called secondary data.

You will write a full report of your investigation. Marks will be awarded for the quality of your report. You should:
- make sure your report is laid out clearly in a sensible order
- use diagrams, tables, charts, and graphs to present information
- take care with your spelling, grammar, and punctuation, and use scientific terms where they are appropriate.

Marks will be awarded under five different headings.

Strategy
- Develop a hypothesis to investigate.
- Choose a procedure and equipment that will give you reliable data.

- Carry out a risk assessment to minimise the risks of your investigation.
- Describe your hypothesis and plan using correct scientific language.

Collecting data
- Carry out preliminary work to decide the range.
- Collect data across a wide enough range.
- Collect enough data and check its reliability.
- Control factors that might affect the results.

Analysis
- Present your data to make clear any patterns in the results.
- Use graphs or charts to indicate the spread of your data.
- Use appropriate calculations such as averages and gradients of graphs.

Evaluation
- Describe and explain how you could improve your method.
- Discuss how repeatable your evidence is, accounting for any outliers.

Review
- Comment, with reasons, on your confidence in the secondary data you have collected.
- Compare the results of your investigation to the secondary data.
- Suggest ways to increase the confidence in your conclusions.

The best advice is 'plan ahead'. Give your work the time it needs and work steadily and evenly over the time you are given. Your deadlines will come all too quickly, especially if you have coursework to do in other subjects.

Creating a case study

Once you have collected the data from your investigation you should look for some secondary data relevant to your hypothesis. This will help you decide how well your data agrees with the findings of other scientists. Your teacher will give you secondary data provided by OCR, but you should look for further sources to help you evaluate the quality of all your data. Other sources of information could include:

- experimental results from other groups in your class or school
- text books
- the Internet.

When will you do this work?

Your school or college will decide when you do your practical investigation. If you do more than one investigation, they will choose the one with the best marks.

Your investigation will be done in class time over a series of lessons.

You may also do some research out of class.

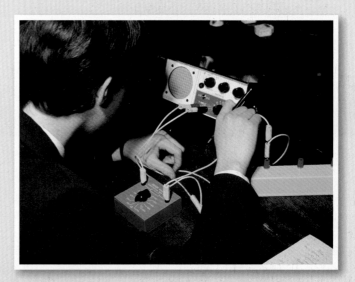

P1 The Earth in the Universe

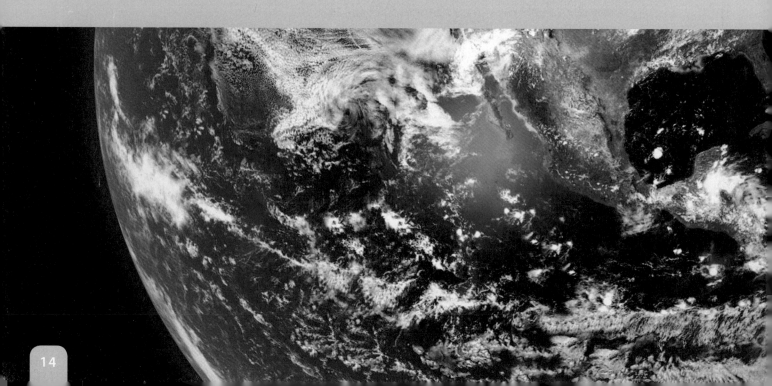

Why study the Earth in the Universe?

Many people want to understand more about the Earth and its place in the Universe. The Earth is a very, very small place in a huge and almost empty Universe. How did the substances we are made of come to be here? What is the history of the Universe itself? Natural disasters, such as earthquakes and volcanic eruptions, can be life-threatening. Why do they happen? Can anything be done to predict them?

What you already know

- The Solar System includes planets, asteroids, minor planets, and comets, all orbiting the Sun.

- Some of the planets have moons orbiting them.

- The Sun is a star, one of a vast number of stars in space.

- The speed of a moving object can be calculated using speed = distance/time.

Find out about

- the history of the Universe

- how scientists develop explanations of the Earth and space

- evidence of the Earth's history found in the rocks

- the movement of the Earth's continents

- what seismic waves can tell us about the Earth.

The Science

Science can explain change. The changes that astronomers observe in stars and galaxies can take millions of years. Stars made the atoms found in everything on Earth, including everything in your body. Closer to home, some changes, such as earthquakes, happen very quickly. Others, such as the formation and disappearance of mountain ranges, take millions of years.

Ideas about Science

Scientists depend on data and careful observations of the Earth and Universe. But scientists need to interpret the data they collect. To do this they must use their imaginations. How are scientific ideas tested? There are often arguments between scientists before new data and explanations are accepted.

Find out about

- ✓ **what is known about the Earth and the Universe**

Our rocky planet was made from the scattered dust of ancient stars. It may or may not be the only place in the whole **Universe** with life.

As the diagrams on these two pages show, scientists know a lot about:
- where and how the **Earth** moves through space
- the history of the Earth.

But there are many things that we still do not know and there are some things we may never know.

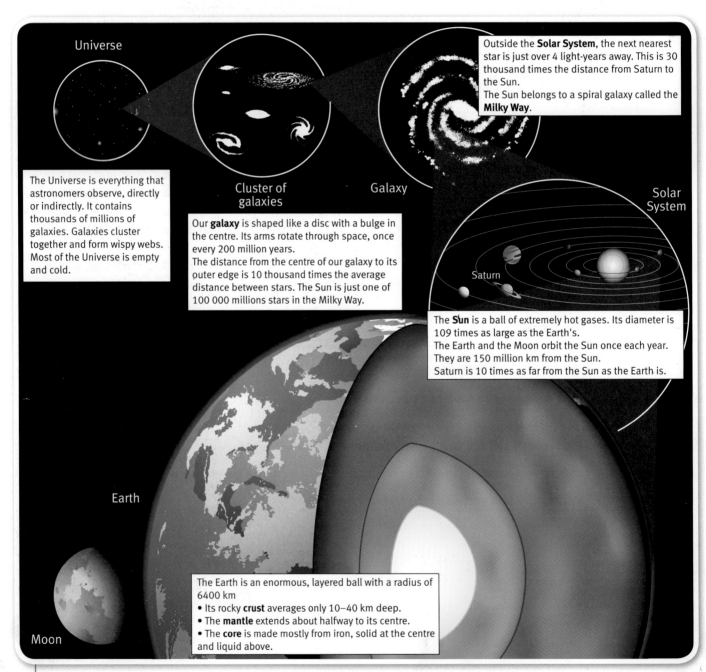

Universe

The Universe is everything that astronomers observe, directly or indirectly. It contains thousands of millions of galaxies. Galaxies cluster together and form wispy webs. Most of the Universe is empty and cold.

Cluster of galaxies

Our **galaxy** is shaped like a disc with a bulge in the centre. Its arms rotate through space, once every 200 million years.
The distance from the centre of our galaxy to its outer edge is 10 thousand times the average distance between stars. The Sun is just one of 100 000 millions stars in the Milky Way.

Galaxy

Outside the **Solar System**, the next nearest star is just over 4 light-years away. This is 30 thousand times the distance from Saturn to the Sun.
The Sun belongs to a spiral galaxy called the **Milky Way**.

Solar System

Saturn

The **Sun** is a ball of extremely hot gases. Its diameter is 109 times as large as the Earth's.
The Earth and the Moon orbit the Sun once each year. They are 150 million km from the Sun.
Saturn is 10 times as far from the Sun as the Earth is.

Earth

Moon

The Earth is an enormous, layered ball with a radius of 6400 km
- Its rocky **crust** averages only 10–40 km deep.
- The **mantle** extends about halfway to its centre.
- The **core** is made mostly from iron, solid at the centre and liquid above.

The Universe is everything there is – from the most distant galaxy to the smallest thing here on Earth.

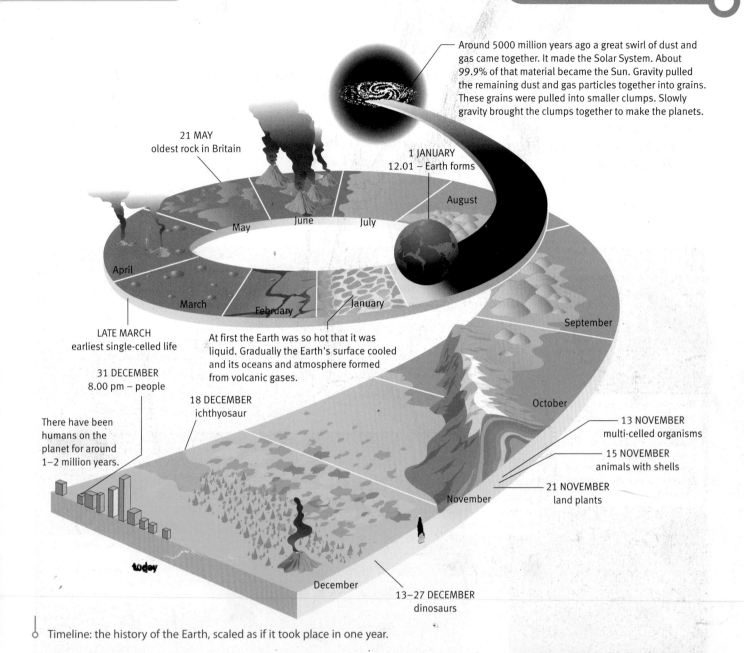

Around 5000 million years ago a great swirl of dust and gas came together. It made the Solar System. About 99.9% of that material became the Sun. Gravity pulled the remaining dust and gas particles together into grains. These grains were pulled into smaller clumps. Slowly gravity brought the clumps together to make the planets.

21 MAY
oldest rock in Britain

1 JANUARY
12.01 – Earth forms

August

June
May
July

April

March
February
January

September

LATE MARCH
earliest single-celled life

At first the Earth was so hot that it was liquid. Gradually the Earth's surface cooled and its oceans and atmosphere formed from volcanic gases.

31 DECEMBER
8.00 pm – people

18 DECEMBER
ichthyosaur

There have been humans on the planet for around 1–2 million years.

October

13 NOVEMBER
multi-celled organisms

15 NOVEMBER
animals with shells

21 NOVEMBER
land plants

November

today

December

13–27 DECEMBER
dinosaurs

Timeline: the history of the Earth, scaled as if it took place in one year.

Questions

1 Using the illustration on the previous page as a source, make a list of seven astronomical objects, in order of size. Start with the Moon and end with the Universe.

2 The timeline above shows the age of the Earth.
  a Redraw it as if it happened over a period of 15 years (roughly your lifetime).
  b On this scale, when did life first appear on Earth? When did the dinosaurs die out?

Key words

- Universe
- Earth
- galaxy
- Solar System
- Milky Way
- Sun
- crust
- mantle
- core

Find out about

- ✔ **what makes up the Solar System**
- ✔ **the process that releases energy in the stars and produces new elements**

Astronomers use telescopes of different sorts to observe the night sky. A telescope gathers radiation, such as light, from distant stars. This radiation carries information that astronomers have learned to decode. This helps us to build up our understanding of the Universe and everything in it.

In some places **light pollution**, dust, and dampness in the air stop radiation getting through the atmosphere.

Solar System

The Earth is part of the Solar System. It is one of eight planets that orbit the Sun. Most of these planets have smaller moons in orbit around them. The **dwarf planets** have similar orbits to the planets. Between Mars and Jupiter is the orbit of the small, rocky **asteroids**.

The orbits of the planets and asteroids are roughly circular. **Comets** are large balls of ice and dust. They have very different orbits, as shown in the diagram below.

Astronomers prefer to work in dark places where they will have a clear view of the night sky. They can analyse light from stars to discover how hot they are and what they are made of.

This diagram of the Solar System shows a typical comet orbit. Comets spend a lot of time far from the Sun. They plunge inwards to pass around the Sun before returning to the cold outer reaches of the Solar System.

The Sun

The Sun is the biggest object in the Solar System. It accounts for 99.9% of the mass. It is also the only object that produces its own light. We see everything else by reflected light.

Scientists once struggled to understand the Sun. It could not be a great ball of fire, because fire needs fuel and oxygen and these would have run out long ago.

Then they found that atoms have a central core, called a **nucleus**. Joining small nuclei together releases energy and creates new elements. This process is called nuclear **fusion**.

Nuclear fusion happens only at extremely high temperatures – millions of degrees. This is the temperature of the inside of a star. Hydrogen nuclei fuse to make helium in the Sun.

Heavy elements are made in stars

The most common elements in the Universe are hydrogen and helium. In stars, hydrogen is fused to make helium. Helium is fused to make heavier elements, such as carbon and oxygen.

At the ends of their lives, some big stars explode as supernovae. Their debris, containing all 92 elements, is scattered through space.

The age of the Solar System

As you saw on page 17, the Solar System formed about 5000 million years ago, from a swirling cloud of dust and gas. This process took millions of years.

As it formed, the Solar System gathered debris from a previous generation of dead stars. Except for hydrogen and helium, the chemical elements that make up everything on Earth come from stars. We are made of stardust.

The remains of a star that exploded as a supernova at the end of its life, spreading out into space. One day, this material may be part of new stars and planets.

An artist's impression of the Solar System as it formed, 5 billion years ago. The Sun's gravity pulled in most of the material. A small amount was left in orbit, which formed the planets, their moons, and everything else.

Questions

1   Put these objects in order of size, from the smallest to the biggest:
    a comet   the Moon   the Sun   an asteroid   the Earth

2   Draw a diagram to compare the orbits of the Earth and a comet around the Sun.

3   Why do scientists believe that there must have been stars that existed and died before the Solar System formed?

Key words

- ✓ **light pollution**
- ✓ **asteroid**
- ✓ **comet**
- ✓ **fusion**
- ✓ **dwarf planet**
- ✓ **nucleus**

Find out about

✔ how astronomers see into the past
✔ ways of measuring the distance to stars

The stars we see in the night sky have a fixed pattern. The Sun, Moon, and planets appear to move against this fixed pattern. This shows that the stars are not part of the Solar System.

Looking back in time

Light moves fast. It could travel the length of Britain in just 6 millionths of a second. At 300 000 km/s, light from the Sun takes just over 8 minutes to reach Earth. This means that you see the Sun as it was 8 minutes ago. You see the stars as they were many years ago.

Although light travels very fast through empty space, it doesn't travel instantaneously from place to place. Its speed is finite.
- Light takes about 1.3 seconds to travel from the Moon to the Earth.
- It takes about 4 years to travel right across the Solar System.
- It takes about 100 000 years to travel right across our galaxy, the Milky Way.

Light-years away

Proxima Centauri is not bright enough to see without a telescope, but it is the closest star outside the Solar System. Light from this star takes 4.22 years to reach Earth. We say that it is 4.22 light-years away.

A **light-year** is a unit of distance used by astronomers. It is the distance travelled by light in one year.

Arcturus is another of the nearer stars. Arcturus is 36.7 light-years away.

How do we know how far away a star is? Astronomers have several methods for measuring the distances to stars. We will look at two of them.

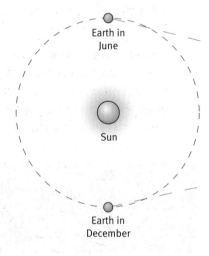

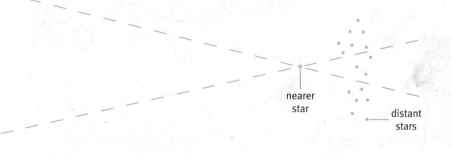

Measuring star distances

Method 1

Parallax: In six months, the Earth moves from one side of the Sun to the other. Seen through a telescope on Earth, a nearby star will shift its position against the background of more distant stars. The nearer a star is, the more it shifts.

This effect is called **parallax**. It provides a way of measuring distance. This method allows astronomers to measure the distance to the nearest stars, including Proxima Centauri.

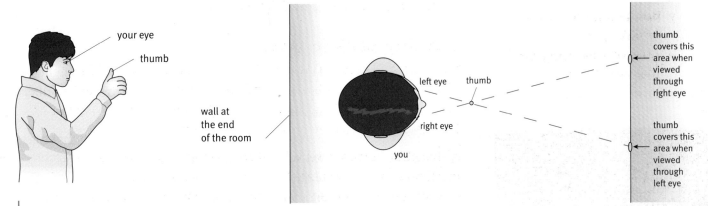

To see the parallax effect, hold up your thumb and look at it with each eye in turn. Your left eye represents the position of the Earth in June, your right eye is the Earth in December.

Method 2

Brightness: The streetlights in the photograph all shine with the same brightness. But the streetlights that are further away appear fainter.

We can use this idea to work out which of two stars is further away. The brighter a star appears to be, the nearer it must be.

But there is a problem with this. We would need to know that the two stars were the same type. Astronomers can analyse the light from the stars to work out how hot they are, and therefore how bright they are. Then we can deduce that, if two stars are the same type, the one that appears fainter must be further away.

Questions

1 It takes light from the Sun eight minutes to reach the Earth.
   a How many seconds is this?
   b Calculate the distance from the Sun to the Earth.

2 Some light from Proxima Centauri is reaching Earth as you read this. How old were you when that light left Proxima Centauri?

3 Suggest why it is easier to use the brightness method of measuring star distances when the observatory is far from any cities or towns.

Key words
✓ light-year
✓ parallax

We know that all streetlights shine with equal brightness, so we can deduce that the ones that appear fainter must be further away.

An artist's impression of the Milky Way galaxy, deduced from observations of many millions of stars. The bright yellow dot in the lower arm represents the position of the Sun, although really the Sun is no brighter than other stars.

Our galaxy

Our galaxy is called the Milky Way. A galaxy is a collection of stars. There are thousands of millions of them, held together by gravity. The Milky Way is called this because it looks like a bright band across the night sky. You need to be in a place where there is little light pollution to see it clearly. It is also easier to see the Milky Way from the southern hemisphere.

For over 2000 years, astronomers had suggested that the Milky Way was made up of large numbers of stars. However, this was not confirmed until 1610, when Galileo first used a telescope to look at the night sky.

By counting the numbers of stars in different directions and estimating their distances, astronomers have built up a picture of the galaxy. The illustration on the left gives an idea of its shape.

- The Milky Way has several arms spiralling out from the centre.
- It has a bulge at the centre.
- The Sun is about half-way out from the centre, in one of the spiral arms.

From clouds to galaxies

In the year 964, a Persian astronomer called Abd al-Rahman al-Sufi published a book of his observations. He was the first person to record a faint smudge of light in the constellation of Andromeda. This came to be described as a **nebula**, the Latin word for a cloud.

Over the centuries telescopes improved and many more nebulae were recorded, but no-one could be sure what they were.

- Were they clouds of gas inside the Milky Way?
- Or were they star clusters, far outside the Milky Way?

In 1925, Edwin Hubble, an American astronomer, used a new telescope to try to find out how far away the Andromeda nebula is. The result was surprising. It seemed it was more than a million light-years away, far outside the Milky Way.

Today's space telescopes show that it is indeed a spiral galaxy, similar to the Milky Way. Its name has been changed to the Andromeda Galaxy.

Billions of billions

We now know that there are thousands of millions of galaxies in the Universe, each made up of thousands of millions of stars – perhaps 10^{22} stars in total.

The Hubble Space Telescope image (below) shows some extremely distant galaxies. The light that made this image left its stars over 10000 million years ago. That was long before the Sun and the rest of the Solar System existed.

Measurement uncertainties

Determining the distances to the stars proved a difficult task for astronomers. First they had to measure the diameter of the Earth's orbit around the Sun. Then they had to use the parallax method to measure the distance to nearby stars. After that astronomers used the brightness method to estimate the distance to further stars. Another method, using variable stars whose brightness varies regularly from day to day, was needed to extend the measurements beyond the Milky Way.

Each of these methods has built-in **assumptions** upon which it depends – for example, the brightness method assumes that two stars that appear similar really are the same type of star.

Each method depended on the results of the previous one, so the final results were, to start with, very **uncertain**. By making many different measurements and comparing the results, astronomers eventually estimated the size of the Universe. And it's very, very big!

A photograph of the Andromeda galaxy. Two other galaxies can be seen in the image: the bright blob above Andromeda and the elliptical smear below it.

An image of distant galaxies, made by light at the end of a long, long journey.

Questions

1 What is the name of the galaxy in which the Solar System is found?

2 What new observation showed that there were objects outside the Milky Way?

3 Roughly how many galaxies are there in the Universe?

4 Explain why measuring the diameter of the Earth's orbit was important for measuring the distance to stars beyond the Solar System.

5 Explain why there are uncertainties about the actual size of galaxies.

Key words

✓ nebula
✓ assumption
✓ uncertain

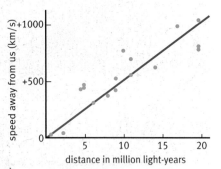

Edwin Hubble published a famous paper in 1929. It showed that more distant galaxies are moving away from us faster.

Imagine yourself on the surface of a very big balloon, looking along a line with galaxies at one-metre intervals. If the balloon is expanding, every metre is growing larger. Let's say the nearest galaxy moves half a metre away from you. In the same time, the second galaxy seems to move away by a metre and the third galaxy by 1.5 metres. The more distant the galaxy, the faster it moves away from you.

The Universe is everything. It is stars and galaxies. It is clouds and oceans. It is bacteria and birds. You are part of the Universe.

A big bang

Until the 20th century, most people thought that the Universe was eternal–it never changed. However, bigger and better telescopes that could see distant galaxies changed all that.

When the astronomers looked at light from distant galaxies, they saw that their spectrum was shifted towards the red end. This is called **redshift**. The amount of redshift shows how fast the galaxies are moving away.

Astronomers had discovered that clusters of galaxies are all moving away from each other. The further they are away, the faster they are moving. The Universe is big and getting bigger. Space itself is expanding.

Scientists now believe that a long time ago the Universe was incredibly hot, tiny, and dense. This explanation is called **big bang** theory.

Testing the theory

The big bang theory passed a major test in 1965. Previously, in 1948, a group of scientists predicted that an afterglow of the big bang event should still fill the whole Universe with microwaves. Years later, two radio engineers in New Jersey noticed an annoying background hiss in their antenna. When they reported this noise, astronomers recognised it as the cosmic microwave background radiation that had been predicted.

The age of the Universe

Picture the expanding Universe – all the galaxies are spreading apart. Now run this imaginary 'film' backwards to see the past history of the Universe – the galaxies all converge at a single point. By working backwards like this, scientists can deduce the age of the Universe.

For their first estimates, 50 years ago, scientists assumed that the galaxies have always moved apart at the same rate as we see today. This may or may not be true. So their answer was only an estimate.

They concluded that the Universe was between 10 000 million and 20 000 million years old. Then, in 2003, new observations of the cosmic microwave background radiation gave a much more precise answer. Scientists estimate that the Universe is 13 700 million years old, plus or minus 200 million years.

Compare this with the age of the Solar System. This was deduced from the age of the oldest rocks on Earth, which are about 4500 million years. The Sun and its planets are about one-third of the age of the Universe.

More evidence for the big bang

Other evidence supports the big bang theory:

- A hot big bang explains why the early Universe was about 76% hydrogen and 24% helium by mass.
- The oldest stars (12 000 million years old) are younger than the Universe. There would be a problem with the theory, if they were older.

Cosmologists at work

Cosmology is the scientific study of the Universe. The big bang is a cosmological theory, devised by cosmologists.

Around the world, there are thousands of cosmologists. Most of them work in universities, usually in groups. When a group develops a new idea, they write a paper for a scientific journal.

Before a scientific paper is published, other experts must first review it. They check it to make sure it has something useful, reliable, and new to say. This process is called **peer review**.

Will the Universe expand for ever?

Cosmologists would like to be able to predict how the Universe will end. The available data is rather uncertain because it is difficult to measure the very large distances to the furthest galaxies, and their speeds.

They would also like to have a better estimate of the mass of the Universe. If the Universe has a high mass, its gravity may cause it eventually to collapse back to a 'big crunch'. However, recent observations of supernovae suggest that the Universe is expanding more and more rapidly. Because the evidence is so uncertain, there are several competing theories of the ultimate fate of the Universe.

Cosmologists make computer models of the Universe. They may have to work with vast amounts of data. This requires a supercomputer, such as the one at Durham University shown here. It can perform 500 billion calculations each second.

Model-independent dark energy test with sigma8 using results from the Wilkinson Microwave Anisotropy Probe

M Kunz, P-S Corasaniti, D Parkinson, and E J Copeland,

Physical Review D **70** 041301 (R) (2004) ICG 04/30

Scientific papers are written for other scientists. You have to be an expert to understand them. Through papers like this, cosmologists share their ideas.

Questions

1 List four observations that support the big bang theory.

2 Explain why the discovery of the cosmic background radiation in 1965 was important to the scientists who had proposed the big bang theory.

Key words

- ✓ big bang
- ✓ redshift
- ✓ peer review

Find out about

- ✓ **James Hutton's explanation for the variety of rocks he found**
- ✓ **how old rocks are and how scientists date them**

Without some way of building new mountains, erosion would wear the continents flat.

Around 250 years ago, people started asking new questions about the history of the Earth. They found fossils of seashells and other marine organisms in rocks at the tops of mountains. 'Why here?' they wondered.

James Hutton and the stories that rocks tell

Rivers carry sediment to the oceans, where it settles at the bottom as sand and silt.

Sediments are compressed and cemented to form sedimentary rocks. In some places, layers of sedimentary rocks are tilted or folded.

James Hutton was a well-educated and observant farmer. He watched heavy rains wash valuable soil off farmers' fields. He also noticed that many rocks are made up of eroded material (now called sedimentary rocks). In his mind, he connected these two observations with the idea of a cycle – continents are both eroded and created.

Using the present to interpret the past

In 1785 Hutton explained his startling new theory of the Earth at a meeting of the Royal Society of Edinburgh. At the time this was like a scientific club. The Society published his theory in its *Transactions*, a kind of newsletter. In this way, his ideas reached a much wider European audience.

What Hutton described is the rock cycle. **Erosion** and deposition of sediment take place, very slowly. Over enormous periods of time, these processes add up to huge changes in the Earth's surface. Erosion makes new soil and is therefore essential to human survival. Heating inside the Earth changes rocks and lifts land up.

The Earth has a history. It was not created all at once. The millions of years over which the Earth has changed are called 'deep time'.

Most Europeans in Hutton's time believed that the Earth had been created exactly as they saw it, just 6000 years earlier. This figure for the Earth's age came from an interpretation of the Christian Bible. They rejected Hutton's theory. It took another century and the support of a leading British geologist, Charles Lyell, before Hutton's ideas became accepted.

Dating rocks

Gradually, geologists learned to work out the history recorded in rocks. They used clues like these:

- Deeper is older – in layered rocks, the youngest rocks are usually on top of older ones.
- Fossils are time markers – many species lived at particular times and later became extinct.
- Cross-cutting features – if one type of rock cuts across another rock type, it is younger. For example, hot magma can fill cracks and solidify as rock.

But these clues only tell you which rocks are older than others. They don't tell you how old the rocks are.

Some rocks are radioactive. Scientists today estimate their age by measuring the radiation that these rocks emit. The Earth's oldest rocks were made about 4000 million years ago.

The development of scientific ideas

This first case study about James Hutton, contains examples of:

- data
- explanations
- the role of imagination.

Data

Fossils, rocks of different types, the way that rock types are layered, folded, or joined.

Explanations

Hutton's idea of a rock cycle, different ways of dating rocks.

Imagination

Hutton could imagine the millions of years needed for familiar processes to slowly change the landscape.

Older rocks tend to lie under younger rocks. Different creatures lived at different times in the past; their fossils can help geologists decide when rocks were formed.

Key word

✓ erosion

Questions

1 In what time order did the creatures shown in the picture of the cliff above live? Which layer has the fallen rock come from?

2 Hutton was able to publish his findings in the *Transactions* of the Royal Society of Edinburgh. Why was this important for him?

Find out about

- ✔ a scientific debate started by Alfred Wegener
- ✔ evidence that the continents are very slowly moving

250 million years ago
Wegener showed how all the continents could once have formed a single continent, called Pangaea.

Key word

- ✔ **continental drift**

Questions

1 In this case study, identify examples of
   a data
   b explanations.

2 'Peer review' involves scientists commenting on the work of other scientists. How did other scientists learn about Wegener's ideas?

3 What were the reasons that other scientists gave for rejecting Wegener's ideas?

How are mountains formed?

A hundred years after Hutton, scientists wanted to know how mountains form. Most geologists believed that the Earth began hot. They compared the Earth with a drying apple, which wrinkles as it shrinks. If the Earth had cooled and shrunk, its surface would have wrinkled too. They claimed that chains of mountains are those wrinkles.

Moving continents?

Scientists discovered radioactivity around 1900. The heating effect of radioactive materials inside the Earth prevents the Earth from cooling. So a new theory of mountain building was needed.

Many people can spot the match between the shapes of South America and Africa. The two continents look like pieces of a jigsaw. Alfred Wegener thought this meant that the continents were moving. They had once been joined together. He looked for evidence, recorded in their rocks.

In 1912 Wegener presented his idea of **continental drift**, and his supporting evidence, to a meeting of the Geological Society of Frankfurt. Geologists around the world read the English translation of his book, *The Origin of Continents and Oceans*, published in 1922.

POLAR EXPLORER DIES

The frozen body of the German meteorologist and polar explorer Alfred Wegener was found on 12 May 1931. Wegener had been leading an expedition in Greenland and went missing just a day after his 50th birthday on 1 November 1930. Unfortunately he is likely to be remembered for being too bold in his science.

Wegener claimed that continents move, by ploughing across the ocean floor. That, he said, explains why there are mountain chains at the edges of continents.

As evidence of continental drift, he found some interesting matches between rocks and fossils on different continents. But most geologists reject such a grand and unlikely explanation for these observations.

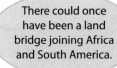

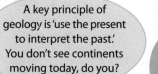

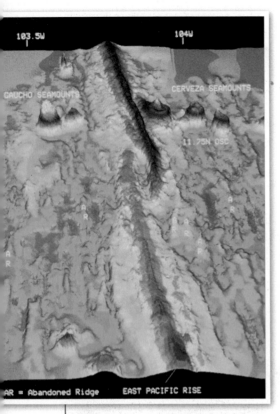

This computer-generated model shows part of the Pacific Ocean floor. (Water is not shown.)

Now and again the Earth's magnetic field reverses, for reasons that scientists still do not fully understand. The magnetic north pole becomes the south pole, and vice versa. Iron-rich rocks record the Earth's field at the time that they solidified.

Mapping the seafloor

During the 1950s the US Navy paid for research at three ocean science research centres. The Navy wanted to know how to:

- use magnetism to detect enemy submarines, and
- move its own submarines near the ocean floor, where they could avoid detection.

A few dozen scientists at these three centres, plus two universities, organis ed many expeditions. They gathered huge amounts of data, and published thousands of scientific papers. Their thinking completely changed our understanding of Earth processes.

From stripes in rocks to seafloor spreading

Scientists started to make maps of the ocean floor. To their great surprise, they found a chain of mountains under most oceans. This is now called an **oceanic ridge**. In 1960 a scientist called Harry Hess suggested that the seafloor moves away from either side of an oceanic ridge. This process, called **seafloor spreading**, could move continents.

Beneath a ridge, material from the Earth's solid mantle rises slowly, like warm toffee. As it approaches the ridge, pressure falls. So some of the material melts to form magma. Movements in the mantle, caused by convection, pull the ridge apart, like two conveyor belts. Hot magma erupts and cools to make new rock.

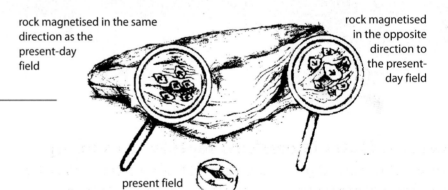

rock magnetised in the same direction as the present-day field

rock magnetised in the opposite direction to the present-day field

present field

A young British research student, Fred Vine, explained the identical stripe pattern found in rock magnetism on either side of two different oceanic ridges. Vine said that if hot magma rises at a ridge and cools to make new rock, then the rock will be magnetised in the direction of the Earth's field at the time. The science journal Nature published his explanation in 1963.

By 1966 an independent group of scientists had found a clearer pattern of symmetrical stripes in magnetic data either side of a third ridge. This forced other scientists to accept the idea of seafloor spreading.

Tanya Atwater was at university studying geology at that time. She describes a meeting of scientists late in 1966. Fred Vine had shown them an especially clear pattern of magnetic stripes.

'[The pattern] made the case for seafloor spreading. It was as if a bolt of lightning had struck me. My hair stood on end. … Most of the scientists [went into that meeting] believing that continents were fixed, but all came out believing that they move.'

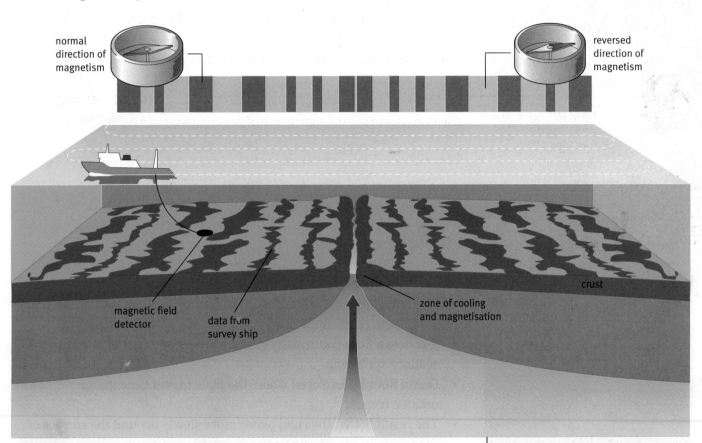

New ocean floor is being made all the time at oceanic ridges. Rock magnetism either side of an oceanic ridge shows the same stripe pattern.

Ocean sediments confirm seafloor spreading

Seafloor drilling in 1969 provided further evidence of seafloor spreading. Sediments further away from oceanic ridges are thicker. This shows that the ocean floor is youngest near oceanic ridges, and oldest far away from ridges.

Questions

4 In this case study, identify examples of:
  a data   b explanations   c predictions

5 Describe carefully how a stripe pattern provides evidence for the seafloor-spreading idea.

Find out about

- **a big explanation for many Earth processes**
- **ways to limit the damage caused by volcanoes and earthquakes**

Each red dot on this map represents an earthquake. Earthquakes happen at the boundaries between tectonic plates.

By 1967, seafloor spreading and several other Earth processes were linked together in one big explanation. It was called plate tectonics.

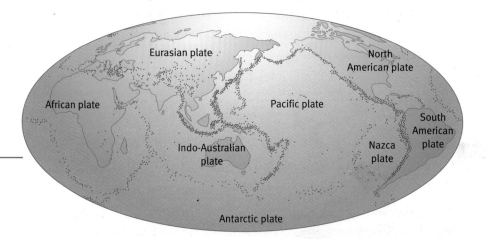

This is the plate tectonics explanation of the Earth's outer layer.

- The Earth's outer layer, or lithosphere, consists of the crust plus the rigid upper mantle.
- It is made up of about a dozen giant slabs of rock, and many smaller ones. These are called **tectonic plates**.
- The lower mantle (below the lithosphere) is hot and soft. It can flow slowly. Currents in the lower mantle carry the plates along.
- The ocean floor continually grows wider at an oceanic ridge by seafloor spreading.
- Ocean floor is destroyed where the plate moves beneath an oceanic trench.
- The result is that the rigid plates move slowly around the surface of the Earth. In places, they move apart. Elsewhere, they push together or slide past one another.

Global Positioning Satellites (GPS) detect the movement of continents. The Atlantic is growing wider by 2.5 cm every year, on average. This is roughly how fast your fingernails grow. In some places, seafloors spread as fast as 20 cm each year.

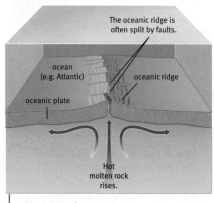

Oceanic ridge.

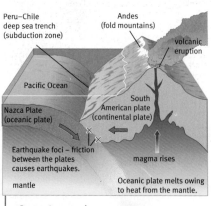

Oceanic trench.

Questions

1. a How far does the Atlantic spread in 100 years (a lifetime)?
   b How far has it spread in 10 000 years (all of human history)?
   c How far has it moved in 100 million years?

2. Today, the Atlantic is about 4500 km wide. How does this compare with your answer to part c?

Plate tectonics explanations

The movement of tectonic plates causes continents to drift.
It also explains:

- parts of the **rock cycle**
- mountain-building
- most earthquakes
- most volcanoes

Earthquakes

In some places tectonic plates slide past each other, as at the San Andreas Fault in California.

The shunting of the Earth's plates causes forces to build up along breaks, called fault lines. Eventually the forces are so great that rocks locked together break, and allow plate movement. The ground shakes, making an **earthquake**.

Earthquakes are common at all moving plate boundaries. The most destructive happen at sliding boundaries on land, or undersea, where they may cause a tsunami.

Making mountains

Collisions between tectonic plates cause mountains to be formed. There are three ways that this can happen.

1   Where an ocean plate dives back down into the Earth, volcanic peaks may form at the surface.

2   The pushing movement at destructive margins can also cause rocks to buckle and fold, forming a **mountain chain**.

3   Sometimes an ocean closes completely, and two continents collide in slow motion. The edges of the continents crumple together and pile up, making mountain chains. This is happening today in the Himalayas and Tibet.

Question

3   Write a step-by-step description of the rock cycle.

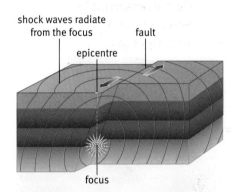

Key words
- ✓ tectonic plates
- ✓ rock cycle
- ✓ earthquake
- ✓ mountain chain

shock waves radiate from the focus
epicentre
fault
focus

The rock cycle

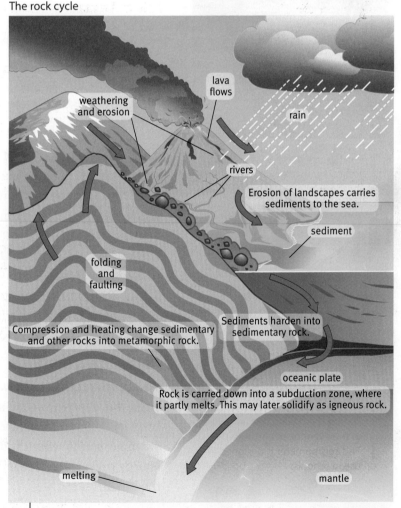

lava flows
weathering and erosion
rain
rivers
Erosion of landscapes carries sediments to the sea.
sediment
folding and faulting
Compression and heating change sedimentary and other rocks into metamorphic rock.
Sediments harden into sedimentary rock.
oceanic plate
Rock is carried down into a subduction zone, where it partly melts. This may later solidify as igneous rock.
melting
mantle

The movement of tectonic plates also plays a part in the rock cycle.

Earthquake damage in Port au Prince, Haiti, following the earthquake of 12 January 2010.

A seismometer – the horizontal and diagonal rods are hinged at the left-hand side, like a garden gate, so that they swing from side to side in the event of an earthquake. This seismometer is designed for use in schools.

Energy of an earthquake

Earth scientists record more than 30 000 earthquakes each year. On average, one of these is hugely destructive.

Vast amounts of energy are released during a powerful earthquake. This energy spreads out from the source of the quake and can be detected at great distances.

Detecting seismic waves

We detect the vibrations of an earthquake using an instrument called a seismometer.

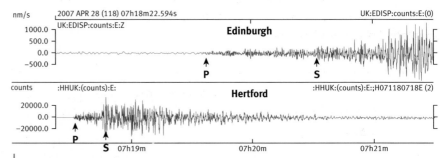

This is a chart, called a seismogram, produced by a seismometer. It shows P-waves and S-waves, which have travelled through the Earth. A seismometer may also detect seismic waves that have travelled around the Earth's surface.

After an earthquake, scientists collect together the data from seismometers in many places around the world to find out as much as possible about the earthquake waves. The picture shows a typical chart, produced when a seismometer detects an earthquake. Two sets of vibrations have been detected:

- The **P-waves** arrive at the detector first (P = primary).
- Then the **S-waves** arrive (S = secondary).

These vibrations that have travelled through the Earth from the site of the earthquake are known as **seismic waves**.

What is a wave?

We are all familiar with waves on the surface of water, but in science the word 'wave' has a special meaning. A **wave** is a repetitive vibration that transfers energy from place to place without transferring matter, for example, sound waves and light waves. In the case of seismic waves, the energy released in an earthquake is carried by the waves around the Earth.

Longitudinal and transverse

P- and S-waves travel differently through the Earth. P-waves travel as a series of compressions (squashed-up regions). In an S-wave, the material of the Earth moves from side to side as the wave travels along.

The pictures show how we can model these two types of wave using a Slinky spring stretched along a table.

To make a P-wave, the end of the spring is pushed back and forth, along the line of the spring. A series of compressions moves along the spring. This type of wave is called a **longitudinal wave**. Sound waves also travel as longitudinal waves.

To make an S-wave, the end of the spring is moved from side to side, at right angles to the line of the spring. This type of wave is called a **transverse wave**. Water waves and waves on a rope are two more examples of transverse waves.

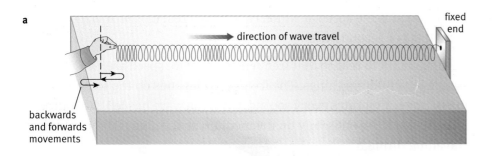

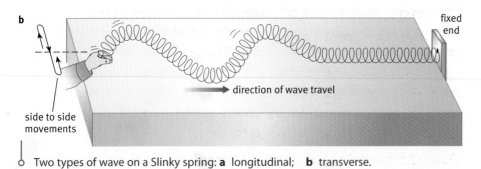

Two types of wave on a Slinky spring: **a** longitudinal; **b** transverse.

Measuring waves

If we draw a simple diagram of a wave, we can define two important quantities:

The **amplitude** is the height of a wave crest above the undisturbed level.

The **wavelength** of the wave is the distance from one crest to the next (or from one trough to the next).

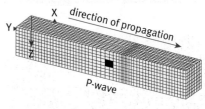

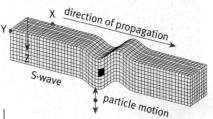

P-waves are sometimes described as 'push-and-pull' waves; S-waves are 'sideways' waves.

Question

1 Look at the seismometer chart on the opposite page. Which waves have a greater amplitude, the P-waves or the S-waves?

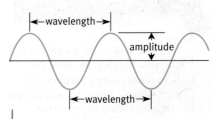

Important information when describing a wave.

2 Using a ruler, measure the drawings of the Slinky springs (on the previous page) as follows:

- To find the wavelength of the longitudinal wave, measure the distance from the centre of one compression to the next.
- To find the wavelength of the transverse wave, measure the distance from the crest of one wave to the next.

3 Which seismic waves have a greater wave speed, P-waves or S-waves? What is the evidence?

4 P-waves travel through the Earth with a wave speed of about 6 km/s. How far will P-waves from an earthquake travel in 5 minutes (300 s)?

Frequency and speed of a wave

Imagine that you are making transverse waves travel along a spring, as in the picture on the previous page. Your hand is the vibrating source of the waves. You can do two things to change the wave.

- Move your hand from side to side by a greater amount. This will give waves with a bigger amplitude.
- Move your hand faster, so that it produces more waves per second. You have increased the frequency of the waves.

The **frequency** of a wave is the number of waves that pass any point each second. It is the same as the number of vibrations per second of the source. Frequency is measured in hertz (Hz). 1 Hz means 1 wave per second.

There is something that you cannot do to the waves on the spring. No matter how you move your hand, you cannot increase their speed. To change the speed of the waves, you would need to use a different spring.

Wave speed is the speed at which each wave crest moves. It is measured in metres per second (m/s).

It is important to realise that frequency and wave speed are two completely different things. The frequency depends on the source – how many times it vibrates every second. Once the wave has left the source, its wave speed depends only on the medium or the material the wave is travelling through.

The distance travelled by a wave can be calculated using the equation:

$$\text{distance} = \text{speed} \times \text{time}$$

Worked example

Measuring distance with a sound wave

The distance-meter emits an ultrasonic sound wave, which bounces from the end of the room and back to the device.

Sound travels at 340 m/s and takes 0.2 seconds to travel the length of the sports hall and back.

How far did the sound travel?

distance = speed × time

distance = 340 m/s × 0.2 s = 68 m

distance = 68 m

How long was the sports hall?

The sound travelled twice the length of the sports hall.

length of sports hall = $\frac{68\,\text{m}}{2}$ = 34 m

The wave equation

Imagine that you are making transverse waves travel along a spring. Moving your hand faster will increase their frequency, as you make more waves per second. The waves look different because their wavelength decreases.

As you know, waves travel out from your hand at a fixed speed. They travel a certain distance in one second. If you make more waves in one second, they will have to be shorter to fit into the same distance.

This shows that there is a link between the frequency of a wave, its wave speed, and its wavelength. Imagine that a source vibrates five times per second. It produces waves with a frequency of 5 Hz. If these have a wavelength of 2 metres in the medium they are travelling through, then every wave moves forward by 10 metres (5 × 2 m) in one second. The wave speed is 10 m/s. In general:

wave speed	=	frequency	×	wavelength
metres per second (m/s)		hertz (Hz)		metres (m)

This link between the three wave quantities applies to all waves of every kind.

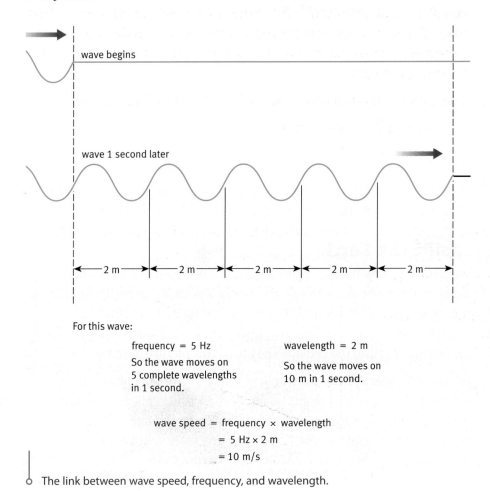

For this wave:

frequency = 5 Hz

So the wave moves on 5 complete wavelengths in 1 second.

wavelength = 2 m

So the wave moves on 10 m in 1 second.

wave speed = frequency × wavelength

= 5 Hz × 2 m

= 10 m/s

The link between wave speed, frequency, and wavelength.

The blue waves have a higher frequency than the red waves and so their wavelength is shorter.

Key words

✓ **frequency**
✓ **wave speed**

Questions

5 A seismic wave travelling through rock has a frequency of 0.5 Hz. The wavelength is 20 km.
   a Calculate the speed of the wave.
   b The wave passes into a different rock where the speed is 14 km/s. What will the new wavelength be?

6 Look at the red and blue waves at the top of the page. Use the diagram to explain what is meant by 'the wavelength of a wave is inversely proportional to its frequency.'

Find out about

- ✔ **how seismic waves travel through the Earth**
- ✔ **how seismic waves reveal the Earth's internal structure**

A geological scan

Seismic waves are a useful tool for geologists who want to look inside the Earth. They don't have to wait for an earthquake to happen; they can set off small explosions on the Earth's surface and detect the reflected waves. The picture shows how this is used in oil exploration.

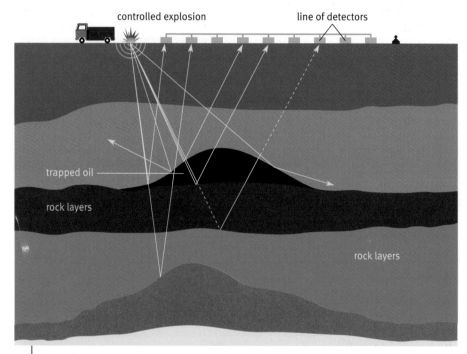

controlled explosion   line of detectors

trapped oil

rock layers

rock layers

Seismic waves spread out from the explosion and are detected by an array of microphones.

From the pattern of reflected waves, the geologists can work out the structure of the underlying rocks. This is similar to the way in which ultrasound waves are used to produce a scan of an unborn baby.

Inside the Earth

Working on a much larger scale, geologists used seismic waves to discover the inner structure of the Earth. The diagram on the left shows the main layers of the Earth. The thin **crust** is on the outside, then the **mantle**, and finally the **core**. It is impossible to dig down more than a few kilometres into the crust, so how was this picture built up?

In the early years of the twentieth century, scientists had set up seismometers at different sites around the world. This allowed them to detect earthquakes that occurred anywhere on the Earth. They could compare the charts from different sites and work out where and when the quake had happened, as well as how strong it was.

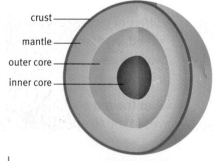

crust

mantle

outer core

inner core

A cross-section through the Earth. This picture is the result of several decades of studies of seismic waves produced by earthquakes.

Shadow zones

Scientists noticed that both P-waves and S-waves reached seismometers close to the earthquake centre, but only P-waves reached seismometers far off, on the other side of the Earth. There was a large 'shadow zone' where S-waves never reached. What could be blocking them?

In 1906, an Irish geologist called Richard Oldham suggested that the Earth had a liquid core. Then, in 1914, a German physicist called Beno Gutenberg published the full answer. He knew that P-waves and S-waves travel differently through the Earth.

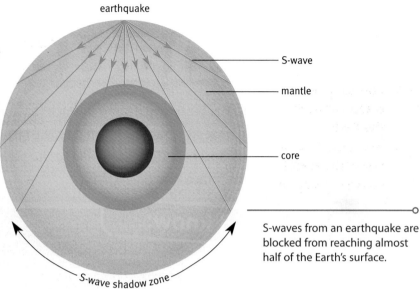

S-waves from an earthquake are blocked from reaching almost half of the Earth's surface.

- P-waves are longitudinal waves and can travel through solids and liquids (just like sound waves).
- S-waves are transverse waves; they can travel through solids but not through liquids.

Gutenberg realised that it was the liquid core that was blocking the passage of S-waves. From the size of the shadow zone, he was able to work out the size of the core. It is about 7000 km thick, roughly half of the Earth's diameter.

As more sensitive seismometers were set up, it became possible to find out more about the Earth's interior. A Danish scientist, Inge Lehmann, looked at the pattern of P-waves. In 1936, she deduced that there must be a small, solid core inside the liquid core.

Questions

1 Name the internal layers of the Earth, starting at the centre.

2 Explain how the pattern of S-waves detected at the Earth's surface showed that the Earth must have a liquid core.

3 Suggest how ideas about the structure of the Earth developed when scientists were working in different parts of the world.

Key words
- ✓ core
- ✓ mantle
- ✓ crust

Science
Explanations

In this module you will see how scientists gather evidence (data and observations). These data and observations are related to the space beyond Earth – the Solar System, stars, and galaxies – and also to the structure of the Earth and the changes that take place in it.

You should know:

- that the Solar System consists of the Sun, eight planets and their moons, dwarf planets, asteroids, and comets
- that the Sun is one of many millions of stars in the Milky Way galaxy
- how the size of the diameters of the Earth, the Sun, and the Milky Way compare
- that light travels at very high speed and the huge distance it travels in one year (a light-year) is used to measure the enormous distances between stars and between galaxies
- that distant objects in the night sky are observed as younger than they are now because the light now reaching us left them a very long time ago
- that the fusion of hydrogen nuclei is the source of the Sun's energy
- that distant galaxies are moving away from us and that this is because the Universe began as a 'big bang' about 14 000 million years ago
- how the ages of the Universe, the Sun, and the Earth compare
- that the Earth is older than its oldest rocks, which are about 4000 million years old
- that Alfred Wegener's theory of continental drift can explain mountain building
- that seafloor spreading, caused by movements in the mantle, provides evidence for continental drift and tectonic plates
- that the movement of tectonic plates causes earthquakes, volcanoes, and mountain building and contributes to the rock cycle
- that earthquakes produce wave motions on and inside the Earth
- that earthquake waves can be transverse or longitudinal
- that the Earth consists of the inner and outer core, mantle, and crust
- that waves are caused by vibrating sources and the number of waves produced each second is the frequency of the wave
- that waves have amplitude and wavelength, where the amplitude is the distance from the top of a crest (or the bottom of a trough) to the undisturbed position, and the wavelength is the length of one complete cycle
- that wave motion can be described by the equations
  distance = speed × time and wave speed = frequency × wavelength

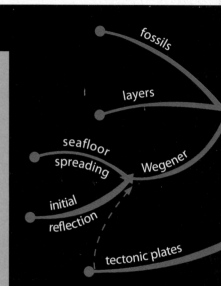

Ideas about Science

In addition to developing an understanding of the structure of the Earth and the nature of the Solar System, stars, and galaxies, you should understand how scientists develop these ideas and how the work of individual scientists becomes accepted or rejected by the scientific community. The case studies in this chapter illustrate these Ideas about Science.

- New scientific data and explanations become more reliable after other scientists have critically evaluated them. This process is called peer review. Scientists communicate with other scientists through conferences, books, and journals.
- Scientists test new data and explanations by trying to repeat experiments and observations that others have reported.

From these you should be able to identify:

- the statements that are data
- statements that are all or part of an explanation
- data or observations that an explanation can account for
- data or observations that don't agree with an explanation.

Scientific explanations should lead to predictions that can be tested. You should know:

- how observations that agree or disagree with a prediction can make scientists more or less confident about an explanation.

Scientists don't always come to the same conclusion about what some data means. The debate about Wegener's idea of continental drift provides an example of this. You should know:

- why Wegener's explanation was rejected at the time
- that some scientific questions have not been answered yet
- distances to many stars and galaxies are not known exactly because they are so difficult to measure
- the ultimate fate of the Universe is difficult to predict.

Review Questions

1 Besides the Sun, the Solar System contains planets, moons, comets, and asteroids.

Explain the differences between these. You should explain how they move and put them in order of size.

2 Vesto Slipher was an astronomer.

In 1915, he measured the speed of a number of distant galaxies.

These are some of his results.

Galaxy	Speed (km/s)
A	1100
B	500
C	1100
D	600
E	300

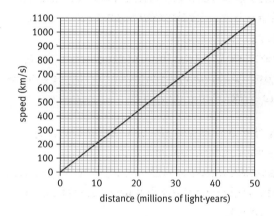

Recent work by astronomers shows that these galaxies are moving relative to us with speed given by the graph:

Use the graph to decide which galaxy in Vesto Slipher's table is 23 million light-years away from us.

3 The following statements describe the events leading to an earthquake. They are in the wrong order. Place them in the right order.

a Great pressure builds up where the plates cannot move easily.

b Two tectonic plates meet at a plate boundary at the San Andreas Fault.

c Slow movements of the magma make tectonic plates move.

d Friction at the edges prevents the plates from sliding easily.

e The sudden movement causes an earthquake.

f When the pressure becomes too great, the plates suddenly slip.

4 Copy and complete the following sentences about observing stars.

Choose words from this list.

detection	galaxies
light	planets
pollution	sound

We can only see stars because they give out ………………………………… .

People in cities find it hard to see stars because of light …………………… .

Astronomers have found that some nearby stars have …………….. in orbit around them.

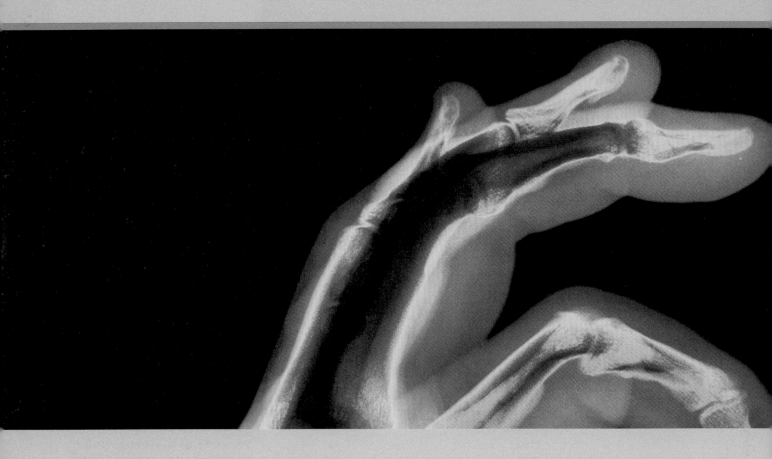

P2 Radiation and life

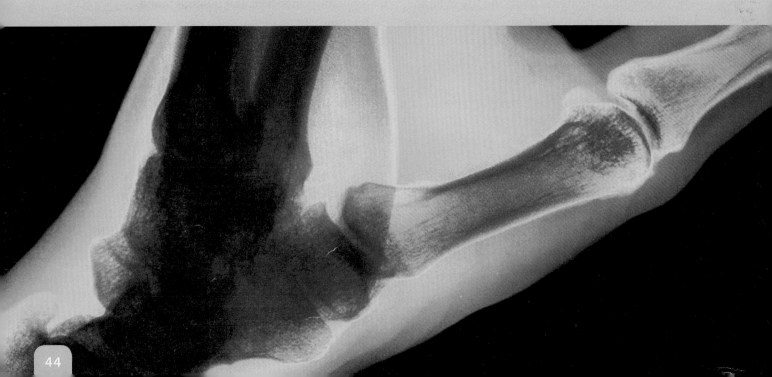

Why study radiation and life?

Human eyes detect one type of radiation – visible light. But there are many other types of 'invisible' radiation. Radiation can be harmful. You hear a lot about the health risks of different radiations. For example, from natural sources such as sunlight, and from devices such as mobile phones. Radiation is involved in climate change, and this is the biggest risk of all.

What you already know

- Light travels in straight lines.
- Light travels very fast.
- When light strikes an object, some may be reflected, some absorbed, and some passes straight through.
- White light can be split into the colours of the spectrum.

Find out about

- how radiation travels and is absorbed
- the evidence of global warming, and its effects
- how information is stored and transmitted digitally
- microwave radiation from mobile phones
- weighing up risks and benefits.

The Science

There are several types of radiation that belong to one 'family', the electromagnetic spectrum. Our communications systems use electromagnetic radiation for transmitting information, such as radio and TV programmes, mobile phones, and computer networks. Science can explain what happens when radiation is absorbed by our bodies. It can also explain how radiation warms the atmosphere, and uses computer modelling to predict global warming.

Ideas about Science

To make sense of media stories about radiation, you need to understand a few things about correlation and cause. How sure can we be that one thing causes another? You can also learn how to evaluate reports from health studies, and how to interpret statements about risk.

Find out about

✓ **benefits and risks of exposure to sunlight**
✓ **the electromagnetic spectrum**

Skin colour

The **ultraviolet radiation (UV)** in sunlight can cause skin cancer. Skin cancer can kill.

Melanin, a brown pigment in skin, provides some protection from UV radiation. People whose ancestors lived in sunnier parts of the world are more likely to have protective brown skin.

What is UV radiation?

Ultraviolet radiation is a member of the 'family' of electromagnetic radiations. Just as visible light can be spread out to form the spectrum from red to violet; electromagnetic radiation can be spread out to form the **electromagnetic spectrum**. UV appears just beyond the violet end of the visible spectrum.

The electromagnetic spectrum is a family of electromagnetic radiations that all travel at the same very high speed, 300 000 km/s, through space.

	radio	microwave	infrared		ultraviolet	X-rays
						gamma rays
lowest frequency				visible		highest frequency

increasing frequency

We can think of electromagnetic radiation travelling as waves. Radio waves are the waves with the lowest frequencies. Gamma rays and X-rays have the highest frequencies.

Sunlight and skin

We need sunlight to fall on our skin. This is because human skin absorbs sunlight to make vitamin D. This nutrient strengthens bones and muscles. It also boosts the immune system, which protects you from infections. Recent research suggests that vitamin D can also prevent the growth and spread of cancers in the breast, colon, ovary, and other organs.

Darker skin makes it harder for the body to make vitamin D. So in regions of the world that are not so sunny there is an advantage in having fair skin. People with dark skin can keep healthy in less sunny countries if they get enough vitamin D from their food.

Balancing risks and benefits

People like sunshine. It can alter your mood chemically and reduce the risk of depression.

But is sunlight good for you? There is no simple answer. Over a lifetime, the risk of developing one type of skin cancer, malignant melanoma, is

Fair skin is good at making vitamin D. But it gives less protection against UV radiation. Melanoma is the worst kind of skin cancer. One severe sunburn in childhood doubles the risk of melanoma in later life.

1 in 91 (UK males) or 1 in 77 (UK females). These figures were calculated in 2009, but the incidence of skin cancer is increasing rapidly. However, if you try and avoid skin cancer by staying indoors, there are risks too.

Protecting your health involves reducing risks, whenever possible, and balancing risks against benefits.

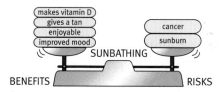

Many people sunbathe. They reckon the benefits outweigh the risks.

Skin cancer warnings ignored

Too much exposure to the Sun is dangerous. A Cancer Research UK survey found a worrying gap between how much people know about skin cancer and how little they actually do to protect themselves in the Sun.

Among 16–24-year-olds, 75% believed that exposure to the Sun might cause skin cancer. But only a quarter of this age group apply high-factor sunscreen as protection. And only a third cover up or seek shade from the Sun.

Key words

- ✓ **electromagnetic spectrum**
- ✓ **ultraviolet radiation (UV)**
- ✓ **factor**
- ✓ **outcome**
- ✓ **correlation**
- ✓ **cause**

Correlation or cause?

A study of 2600 people found that people who were exposed to high levels of sunlight were up to four times more likely to develop a cataract (clouding of the eye lens). Exposure to sunlight is possibly a **factor** in causing cataracts. Eye cataracts are an **outcome**. There is a **correlation** between exposure to sunlight and eye cataracts. But doctors do not say that exposure to sunlight will produce cataracts – it may not be the **cause**. There are other risk factors involved, such as age and diet.

Questions

1 Look at the diagram of the electromagnetic spectrum on the opposite page.
   a Which type of radiation has the lowest frequency?
   b Which colour of visible light has the highest frequency?

2 A person with dark skin moves to live in a region where there are few sunny days. Why should they try to spend a lot of time out of doors?

3 Exposure to sunlight increases your risk of developing skin cancer. List some benefits of staying indoors and avoiding direct sunlight. Suggest some risks of staying indoors.

4 Look at the information about the risk of melanoma. Who is more at risk of developing skin cancer? Suggest a reason why.

Find out about

- ✔ **sources of light, and the paths light follows**
- ✔ **how the ozone layer protects life on Earth**

Key words

- ✔ **source**
- ✔ **emits**
- ✔ **transmit**
- ✔ **reflect**
- ✔ **absorb**
- ✔ **ozone layer**
- ✔ **atmosphere**
- ✔ **CFCs**

Coloured materials added to glass can absorb some colours of light and transmit others.

A beautiful world

All radiation has a **source** that **emits** it. Then it has a journey. It spreads out, or 'radiates'. Radiation never stands still. Some radiation, at the end of its journey, causes chemical changes at the back of your eye. That radiation is visible light.

Some materials, like air, are good at **transmitting** light. They are clear, or transparent. On the way from the source to your eyes light can be **reflected** by other materials. The objects around you would be invisible if they did not reflect light.

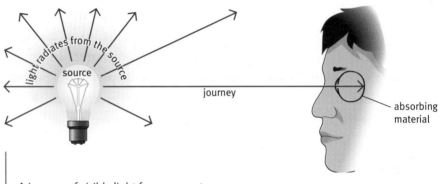

A journey of visible light from source to eye.

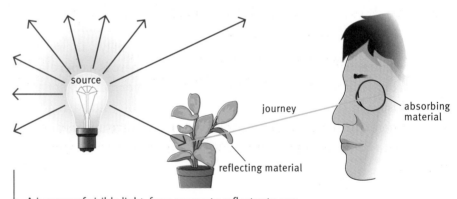

A journey of visible light, from source to reflector to eye.

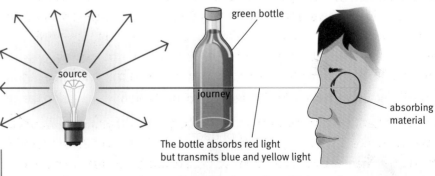

A journey from source to detector, but with absorption of light on the way.

An absorbing atmosphere

Radiation from the Sun contains infrared radiation, visible light, and ultraviolet radiation. Some of this radiation is transmitted through the atmosphere so that it reaches the ground. Fortunately for us, most of the harmful UV radiation is **absorbed** as the Sun's radiation passes downwards through the atmosphere. Life on Earth depends on the ozone layer absorbing UV radiation.

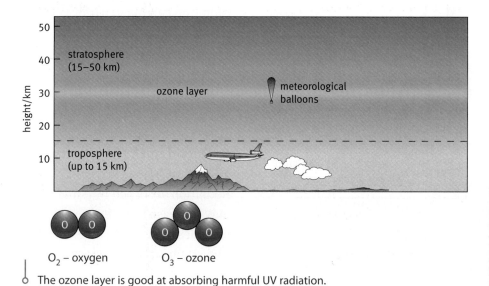

The ozone layer is good at absorbing harmful UV radiation.

The **atmosphere** is a mixture of gases, including oxygen. In the upper atmosphere some of the oxygen atoms combine in threes to make ozone. It makes an **ozone layer**.

The ozone layer is good at absorbing UV radiation. When UV radiation is absorbed its energy can break ozone and oxygen molecules, making free atoms of oxygen.

These chemical changes are reversible. Free atoms of oxygen in the ozone layer are constantly combining with oxygen molecules to make new ozone.

Ozone holes

Humans have created a problem. Some synthetic (man-made) chemicals, such as **CFCs** (chlorofluorocarbons) used in fridges, have been escaping into the atmosphere. They turn ozone back into ordinary oxygen. So more UV radiation reaches the Earth's surface. This happens strongly over the North and South Poles. More UV radiation can reach the Earth's surface through the 'hole in the ozone layer'.

The international community is now dealing with the problem. Aerosol cans can no longer use CFCs. Old fridges have the CFCs carefully removed at the end of their working life. However, the ozone layer may take decades to return to its original thickness.

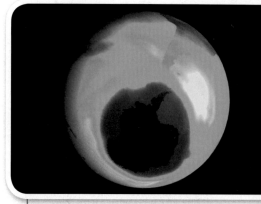

This image has been made by sensing ozone. Dark colours represent less dense ozone. There seems to be a 'hole' in the protective layer.

Old fridges waiting to have CFCs removed.

Questions

1 Can glass reflect light? Explain how you know.

2 Materials can transmit, reflect or absorb light. Which one of these is glass best at? How can this property of glass be changed?

3 Use the words – *source*, *reflect*, and *detector* to explain how you can see to read at night, using a torch.

4 Why is it important for life on Earth that the atmosphere absorbs most of the UV radiation that comes from the Sun?

When materials absorb electromagnetic radiation they gain energy. Exactly what happens depends on the type of radiation.

Find out about

- **what happens when electromagnetic radiation is absorbed**
- **why some kinds of radiation are more dangerous than others**

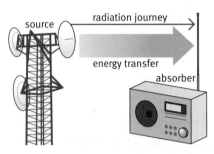

Metal aerials can absorb radio and microwave radiation. The process creates electrical patterns inside the metal.

Radiation can cause a varying electric current in a metal wire

Patterns of microwave and radio radiation can make patterns of electric current in radio aerials.

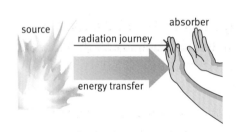

A fire transfers energy to the world around it. Surfaces in its surroundings, including people, absorb the radiation and gain the energy.

Radiation can have a heating effect

Radiation absorbed by a material may increase the vibration of its particles (atoms and molecules). The material gets warmer.

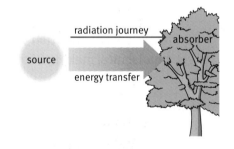

A leaf takes energy from the Sun's radiation so that photosynthesis can happen.

Radiation can cause chemical changes

If the radiation carries enough energy, the molecules that absorb it become more likely to react chemically. This is what happens, for example, in photosynthesis, and in the retinas of your eyes.

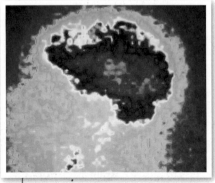

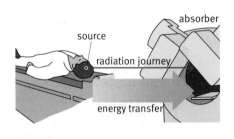

This medical image was made by a gamma camera. Each dot on the image was made by a single ionisation event.

Ionisation can damage living cells

If the radiation carries a large amount of energy, it can remove an electron from an atom or molecule, creating charged particles called **ions**. This process is called **ionisation**. The ions can then take part in other chemical reactions. Ionisation can damage living cells.

Radiation arrives in energy packets

Radiation is 'grainy'. It carries energy in small packets called **photons**:

- Sources emit energy photon by photon.
- Absorbers gain the energy photon by photon.

The energy deposited by a beam of electromagnetic radiation depends on both:

- the number of photons arriving at the absorber
- and the energy that each photon delivers.

Ionising and non-ionising radiation

Look back at the electromagnetic spectrum on page 46. Radiations with the highest frequencies have the photons with the highest energies. **X-ray** photons and **gamma ray** photons carry most energy. Radio photons carry least energy.

Sources of gamma rays, X-rays, and high-energy UV radiation pack a lot of energy into each photon. A single photon has enough energy to ionise an atom or molecule, by knocking off an electron. Gamma rays, X-rays, and some UV are **ionising radiations**.

Visible, infrared, microwave and radio radiations are all **non-ionising radiations** because a single photon does not have enough energy to ionise an atom or molecule. The main effect of these radiations is warming. The lower the photon energy is, the smaller the heating effect of each photon.

Lying in the sunshine, infrared, and visible radiations have a warming effect; UV radiation can initiate a chemical change that could start skin cancer.

Key words

- ✓ ions
- ✓ ionisation
- ✓ photons
- ✓ ionising radiation
- ✓ non-ionising radiation
- ✓ X-rays
- ✓ gamma rays

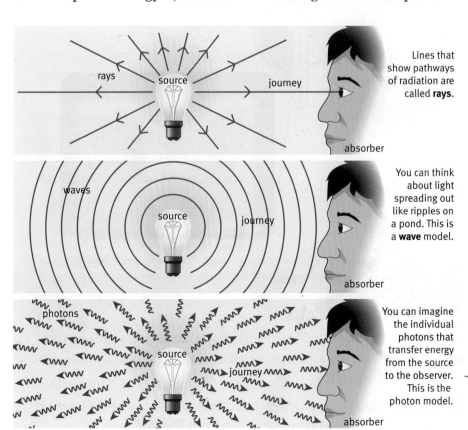

Lines that show pathways of radiation are called **rays**.

You can think about light spreading out like ripples on a pond. This is a **wave** model.

You can imagine the individual photons that transfer energy from the source to the observer. This is the **photon** model.

Question

1 There is radio radiation passing through your body right now.
   a Where does the radio radiation come from?
   b Why does it not have any ionising effect?
   c Does it have a heating effect? Explain.

Radiation transfers energy. There are different ways of thinking about how it travels between source and absorber.

Find out about

- ✔ **reducing the risk from ionising radiation**
- ✔ **how ionising radiation can affect body cells**

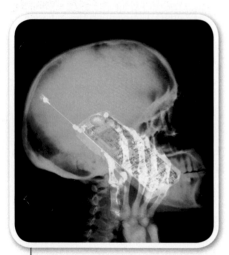

Both the health benefits and the risks of X-rays are well known. Using a mobile phone has benefits but uncertain risks.

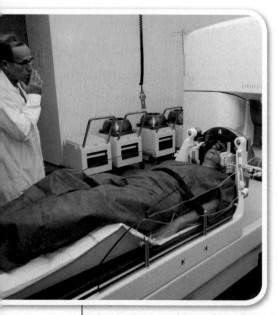

This patient is undergoing gamma radiotherapy; gamma radiation from the machine above him is directed towards a cancerous tumour in his body.

Using X-rays

X-rays were discovered in the 1890s. They soon caught on as a medical tool, and they have saved many lives. How do they work?

X-rays are produced electrically, in an X-ray tube. A beam of X-rays is shone through the patient and detected on the other side using an X-ray camera.

As the beam passes through the patient, it is partly absorbed. Bone is a stronger absorber than flesh, and so bones show up as 'shadows' on the final image. We say that the **intensity** of the beam is reduced as it is absorbed.

The intensity of a beam of electromagnetic radiation is the energy that arrives at a square metre of surface each second. This tells us how 'strong' the beam is.

The intensity of any radiation is less if you are further away from the source. This is because the radiation spreads out over a wider and wider area.

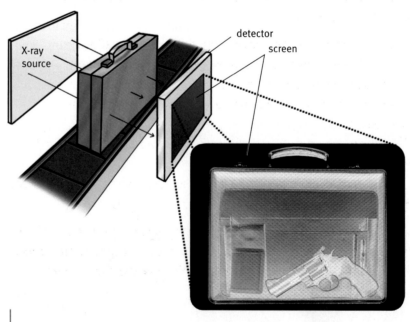

X-ray machines are used in airport security checks. Like bone, metal objects absorb X-rays strongly and so reduce the intensity of the beam.

Using gamma rays

Gamma radiation is similar to X-rays but it comes from **radioactive materials**. These are substances that emit radiation all the time – you can't switch them off.

Gamma radiation is used, like X-rays, for imaging a patient's internal organs. It is also used to destroy cancer cells.

Discovery of a correlation

Alice Stewart (see photo) and George Kneale carried out a survey on a large number of women and their children. They discovered a correlation between X-ray exposure of mothers during pregnancy and cancers in their children.

There is a plausible **mechanism** that could explain this correlation. X-ray photons can ionise molecules in your body, and is particularly risky if DNA molecules are affected. This can disrupt the chemistry of body cells, and cause cancer. So the link is more than just a correlation. X-rays can, in a few cases, cause cancer.

This study made doctors more cautious about using X-rays. The radiation is more damaging to cells that are rapidly dividing, such as a growing fetus or a small child. So the risks associated with X-rays for small children and pregnant women usually outweigh any benefit.

Reducing the risk

When a patient has an X-ray, the equipment and procedures are designed to keep the X-ray exposure to the minimum that still produces a good image. People who work with ionising radiation must also be protected from its effects.

There are several ways to reduce exposure to ionising radiation:

- **time:** the shorter the time of exposure, the less radiation is absorbed so the smaller the chance of damage to cells
- **distance:** intensity decreases as radiation spreads out from the source
- **shielding:** use materials such as lead and concrete, which absorb radiation strongly
- **sensitivity:** use a detector that is more sensitive so that less radiation is needed to produce an image.

UV photons have enough energy to change atoms and molecules, which can initiate chemical reactions in body cells. This can cause skin cancer. This is why it is advisable to cover up with clothes and sunscreen to reduce the risk on a sunny day.

Obituaries

Alice Stewart

Alice Stewart was a British doctor. She collected and analysed information from women whose children had died of cancer between 1953 and 1955. Soon the answer was clear. On average, one medical X-ray for a pregnant woman was enough to double the risk of early cancer for her child.

Key words

- ✓ intensity
- ✓ radioactive materials
- ✓ mechanism

Questions

1 a Name the three types of electromagnetic radiation that are also ionising radiations.
  b Which of these has the least energetic photons?

2 In the article about Alice Stewart, what is the **outcome** she is studying? What is the **factor** that might be causing this outcome?

3 a Why is the link between X-ray exposure during pregnancy and childhood cancer believed to be a 'cause' and not just a 'correlation'?
  b Why do doctors still use X-rays, despite this link?

Find out about

- ✓ the heating effect of microwaves
- ✓ how microwave ovens work safely
- ✓ radiation from mobile phones

Microwave ovens

A microwave oven uses **microwave radiation** to transfer energy to absorbing materials. Once the materials have absorbed the energy of the radiation, it ceases to exist as radiation. Molecules of water, fat, and sugar are good absorbers of microwave radiation. When they absorb microwave radiation, they vibrate strongly; in other words, their temperature rises.

A potato contains water, so it absorbs microwave radiation. The intensity of the radiation decreases as it passes into the potato. The particles of glass or crockery absorb very little energy from the radiation. It does not increase their vibrations, so bowls and mugs are not heated directly. They are heated by the hot food or drink inside them.

Absorption of energy by a potato. The radiation is gradually absorbed as it passes into the potato. It may not reach the middle of a very large potato. The plate absorbs very little energy from the microwave radiation.

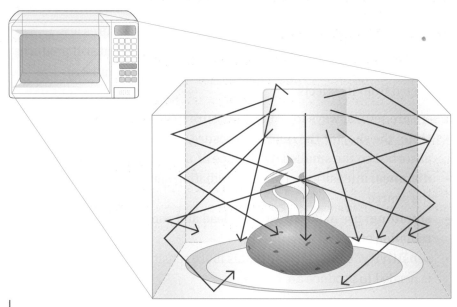

Transmit, absorb, reflect … Inside a microwave oven, materials like glass and pottery only partially transmit the radiation. The metal walls reflect it. Some substances, such as the water in a potato, absorb it.

Safety features

People contain water and fat, two absorbers of microwave radiation. So microwaves could cook you. The oven door has a metal grid to reflect the radiation back inside the oven. A hidden switch prevents the oven from operating with its door open.

How well cooked?

Any material that absorbs non-ionising radiation (such as microwaves) gets hot. The heating effect depends on the **intensity** of the radiation and its **duration** (the exposure time).

Key words

- ✓ microwave radiation
- ✓ intensity
- ✓ duration
- ✓ infrared
- ✓ principal frequency

You control the amount of cooking in a microwave oven by adjusting:

- the power setting (to control the intensity)
- the cooking time (to control the duration of exposure).

Cooked brain?

Mobile phones use microwave radiation to send signals back and forth to the nearby phone mast. When you make a call, the fairly thick bones of your skull absorb some of this radiation. But some reaches your brain and warms it, ever so slightly. Vigorous exercise has a greater heating effect.

There is no evidence that this exposure is harmful. However, some people use a hands-free kit so that the phone is further from their head. The radiation is less intense because it spreads outwards from the phone.

We're all radiators

Hot objects glow brightly. They emit visible light. In fact, cool objects emit electromagnetic radiation too. This is invisible **infrared** radiation. Even objects whose temperature is far below 0°C emit weak infrared radiation. This has a very low frequency.

The hotter an object is, the higher the frequency of the radiation it emits. Very hot objects, such as the hottest stars, emit radiation whose **principal** or main frequency is in the ultraviolet region of the electromagnetic spectrum.

A mobile phone stops radiating when you stop speaking. It also sends a weaker signal when you are close to the phone mast. That's to save the battery, but it also means that less radiation penetrates your head.

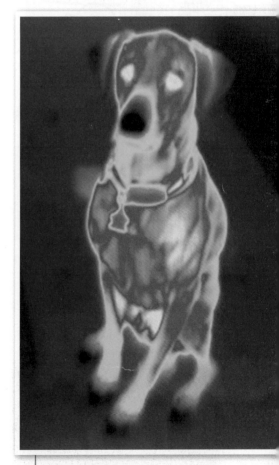

Animals (including people) are too cool to emit visible light, but a special camera can detect the infrared radiation they give out.

Questions

1 What radiations are on either side of microwave radiation in the electromagnetic spectrum?

2 Why doesn't microwave radiation cause ionisation?

3 Why is it important that the walls and door of a microwave oven reflect the microwave radiation?

4 Many people imagine that it is dangerous to live close to a mobile phone mast. Why might your exposure to microwave radiation from your phone be reduced if you lived close to a mast?

5 The damage caused by radiation depends on the energy of the photons. Energy of photons of electromagnetic radiation is *proportional to* the frequency of the radiation. Explain what is meant by 'proportional to.'

Data collected by the meteorological Office shows that Bognor Regis gets more hours of sunshine than any other town in England. This data is more reliable than any individual's recollections.

Are summers now hotter and winters milder than they once were? This is a question about **climate**, or average weather in a region over many years. You cannot answer it from personal experience, because you can only be in one place at a time. And memory can be unreliable. Instead, you need to collect and analyse lots of data.

A comfortable temperature for life

The Earth's average temperature is about 15 °C, which is very comfortable for life. Why does it have this temperature?

Firstly, we are in orbit around the Sun. The Earth's surface absorbs radiation from the Sun, and this warms the Earth. At the same time, the Earth emits radiation back into space.

- During the day, our part of the Earth is facing the Sun. The Sun's radiation is absorbed by the Earth. It warms us up and the temperature rises.
- At night, we are facing away from the Sun. Energy radiates away into space and the Earth gets colder.
- The Earth is cooler than the Sun, so the radiation it emits has a lower principal frequency. It can be absorbed by the atmosphere.

Without the Sun to keep topping up the Earth's store of energy, the Earth's temperature would soon fall to the temperature of deep space, about −270 °C.

The greenhouse effect

Without its atmosphere, the Earth's average surface temperature would be –18 °C. That's how cold it is on the Moon. This warming of the Earth by its atmosphere is called the **greenhouse effect**.

Life on Earth depends on the greenhouse effect. Without it, the Earth's water would be frozen. Water in its liquid form is essential to life.

Greenhouse gases

If the atmosphere consisted entirely of the commonest gases (nitrogen and oxygen), there would be no greenhouse effect.

Questions

1 Personal experience does not provide reliable evidence of climate change. Why not?

2 Which of the following gases found in the Earth's atmosphere are *not* greenhouse gases?: nitrogen, methane, oxygen, carbon dioxide, water vapour?

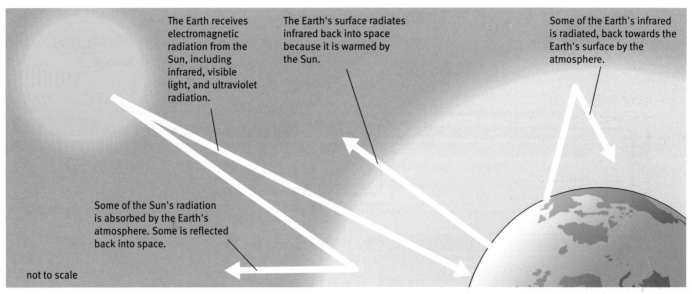

The Earth receives electromagnetic radiation from the Sun, including infrared, visible light, and ultraviolet radiation.

The Earth's surface radiates infrared back into space because it is warmed by the Sun.

Some of the Earth's infrared is radiated, back towards the Earth's surface by the atmosphere.

Some of the Sun's radiation is absorbed by the Earth's atmosphere. Some is reflected back into space.

not to scale

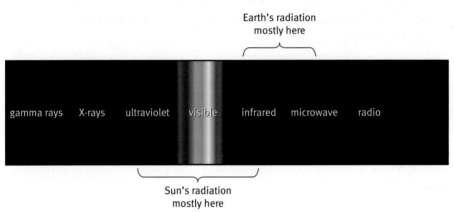

Earth's radiation mostly here

gamma rays   X-rays   ultraviolet   visible   infrared   microwave   radio

Sun's radiation mostly here

There is an energy balance between radiation coming in and going out of the atmosphere. The atmosphere lets in infrared radiation from the Sun, but prevents the infrared emitted by the Earth from escaping. This is because the Earth's radiation has lower frequencies than the Sun's, and these are absorbed by the atmosphere.

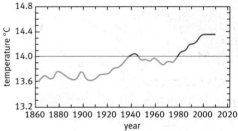

The Earth's surface temperature has risen over the past 150 years (data from weather stations).

Tiny amounts of a few other gases make all the difference. Carbon dioxide, methane and water vapour absorb some of the Earth's infrared radiation. They are called **greenhouse gases**.

Past temperatures

Weather stations have kept temperature records for over a century. The graph opposite shows the results.

There is a clear pattern. The Earth's average temperature has been rising since 1800. This conclusion is supported by evidence from Nature's own records (growth rings in trees, ocean sediments, air trapped in ancient ice).

Most climate scientists think that carbon dioxide (CO_2) in the atmosphere is causing the rise in temperatures. Why?

- Temperatures and CO_2 levels have risen at the same time.
- Evidence from the distant past suggests that temperature and CO_2 levels go up and down together.
- Scientists can explain how CO_2 in the atmosphere absorbs radiation and raises the temperature.

Questions

3 All of the statements about CO_2 and the Earth's average temperature describe correlations. Which statement is also about cause and effect?

4 Look at the temperature graph. Use the graph to describe how the Earth's surface temperature has changed over the past 150 years.

The carbon cycle

Carbon dioxide is a greenhouse gas that plays a key role in global warming. Industrial societies produce CO_2 as never before.

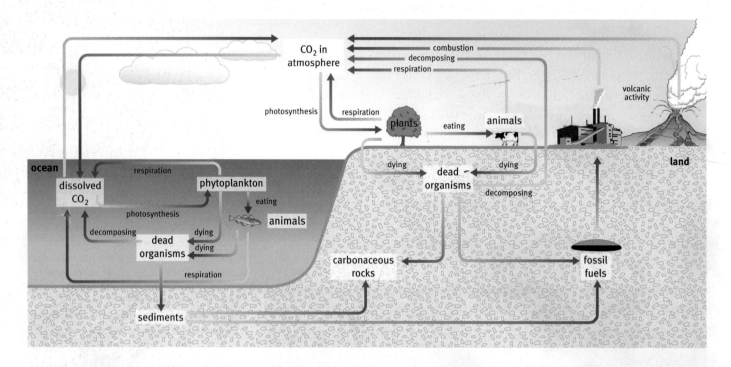

The Earth's crust, oceans, atmosphere, and living organisms all contain carbon. Carbon atoms are used over and over again in natural processes. The **carbon cycle** describes the stores of carbon and processes that move carbon.

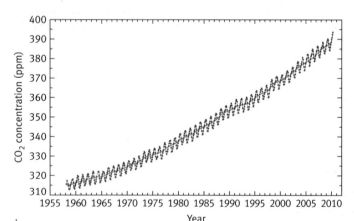

Carbon dioxide concentrations have been recorded at Mauna Loa in Hawaii since 1958. They rise and fall each year, but the overall trend has been an increase of about 1.5 ppm per year since 1980.

Carbon dioxide in the atmosphere

Hundreds of millions of years ago, the amount of CO_2 in the atmosphere was much higher than it is today. Green plants made use of CO_2 and released oxygen. This made life possible for animals. Eventually, lots of carbon was locked up underground in the form of fossil fuels, as well as carbonaceous rocks such as limestone and chalk.

For thousands of years CO_2 levels were stable. Plants absorbed CO_2 during photosynthesis, then animals and decomposers returned it to the atmosphere during respiration. Humans burned wood, but that was balanced by new trees growing. Two hundred years ago, the industrial revolution changed all that as fossil fuels were burned in increasing quantities.

Human activities release carbon

People want to live comfortably. In some parts of the world, many feel they have a right to processed foods, unlimited clean water and electricity, manufactured goods, and bigger houses and cars. All of these things require energy.

But whenever fossil fuels (coal, oil, and gas) are burned, they increase the amount of carbon dioxide in the atmosphere. Methane, another greenhouse gas, is produced by grazing animals and from rice paddies. Cutting down or burning forests (**deforestation**) also releases CO_2, and reduces the amount removed by photosynthesis.

Although methane is the more effective greenhouse gas, carbon dioxide produced by human activities has a bigger effect. This is because the amount of CO_2 is so huge – thousands of millions of tonnes each year. This is why there are international agreements on reducing 'carbon emissions'.

Motor vehicles are a major source of greenhouse gas emissions.

This power station supplies enough electricity for a major city. Every day it uses several trainloads of coal and sends thousands of tonnes of carbon dioxide into the atmosphere.

Air transport is a big user of fossil fuels. Aviation fuel is cheap because it is untaxed, unlike petrol for cars.

People in the UK use more energy for keeping buildings warm than for anything else.

Questions

5 Study the diagram of the carbon cycle opposite.
   a List six processes that release CO_2 into the atmosphere.
   b List two processes that remove CO_2 from the atmosphere.
   c Which of the above processes has changed so that the amount of CO_2 in the atmosphere is increasing?

6 Look at the graph of CO_2 levels opposite. Explain its shape: why does it go up and down every year? Why is the long-term trend upwards?

7 Forest land can be cleared for farming by burning trees. Explain why tree burning increases the amount of carbon dioxide in the atmosphere.

Key words
   ✔ carbon cycle
   ✔ deforestation

The tiny bubbles in this slice of ancient ice from Antarctica contain air trapped hundreds of thousands of years ago.

An 'ice core' like this stores a record of the Earth's changing atmosphere over hundreds of thousands of years.

Nature's records

The polar ice caps are frozen records of the past. In parts of the Antarctic, ice made from annual layers of snow is four kilometres thick. That ice contains tiny bubbles of air, a record of the atmosphere over 800 000 years. It shows that climate has always changed. There have been ice ages and warm periods.

But temperatures have never increased so fast as during the past 50 years.

Natural factors change climates

Over the long term, natural factors cause climate change. For example:

- the Earth's orbit changes the distance to the Sun by small amounts
- the amount of radiation from the Sun changes in cycles
- volcanic eruptions increase atmospheric CO$_2$ levels.

These factors cause much slower changes than we are seeing today, but they still must be taken into account when scientists try to determine whether human activities are causing the climate to change.

Climate modelling

The atmosphere and oceans control climates. Climate scientists use **computer models** to predict the effects of increasing CO$_2$ levels.

What is a computer model? Climate models are similar to the models used for day-to-day weather forecasting. They use everything climate scientists know about how the atmosphere and oceans behave. For example, we know that, at present, the Earth absorbs about 1% more energy from the Sun than it radiates back into space. This extra energy warms the oceans, increasing the rate at which water evaporates, and the amount of water vapour in the atmosphere. The computer models can calculate how this will affect temperatures around the world.

Computer models are tested using data about the Earth's climate in the past. If they can correctly account for this data, it is more likely that their predictions for the future will be accurate.

However, the further we try to look into the future, the greater the uncertainty in our predictions.

Alarming predictions

What these models show is alarming.

- Human activities now contribute more to climate change than natural factors.

- Future emissions of greenhouse gases are likely to raise global temperatures by between 2 and 6 °C during your lifetime.
- If CO_2 concentration rises much further, climate change may become irreversible.
- To stabilise climates, carbon emissions would need to be reduced by at least 70% globally.
- In the UK, winters will become wetter and summers drier.

Climate change is a slow process. It may take 20 to 30 years for climates to react to the extra CO_2 already in the atmosphere. So global temperatures are guaranteed to rise by 2 °C. Ice will continue to melt, and sea levels continue to rise, for the next 300 years or so, even if humans today stopped producing any CO_2 at all.

Effects of climate change

Human societies depend on stable climates. The risks associated with global warming are enormous.

Extreme weather: There are likely to be more extreme weather events (violent storms, heat-waves). This is because higher temperatures will cause greater convection in the atmosphere, and more evaporation of water from both oceans and the land.

Rising sea levels: Water in the oceans will expand as it gets hotter, so sea level will rise. In addition, continental ice sheets, such as in Antarctica and Greenland, may melt, adding to the volume of the oceans. Low-lying land will be flooded, causing a particular problem for people who live in river deltas such as the Ganges delta in Bangladesh, or on low-lying islands.

Drought and desertification: Reduced rainfall may make it impossible to grow staple crops in some areas. Tropical areas may become drier, leading to the expansion of deserts such as the Sahara.

Health: Malaria will spread if mosquitoes can breed in more places.

Climate-change sceptics

Thousands of climate scientists have contributed to our understanding of how human activities are affecting the climate. They publish their results in scientific journals and test each other's ideas.

Some scientists and many other non-specialists have challenged aspects of this work. These people are sometimes called 'climate sceptics' or 'deniers'. Despite these challenges new evidence usually supports climate scientists' ideas.

These maps, based on satellite photographs, show how the area of the Arctic ice sheet decreased between 1980 and 2007.

Questions

1 Sea levels are rising. Give two reasons why they are likely to continue rising in the future.

2 Explain why rising temperatures will give rise to more violent storms.

3 Which **two** effects of global warming may make it difficult to grow some food crops in particular regions?

4 Make a list of the scientific uncertainties mentioned on these pages.

Paying the price

The world's poorest countries will be least able to deal with the effects of climate change. Their people are the most vulnerable. Yet it is the wealthier countries that are responsible for the problem.

The bar chart shows this:

- The height of each block in this graph shows how much CO_2 is emitted per person each year – North Americans emit most.
- The area of each block shows the total emissions for each region – people in Europe and North America are responsible for over 60% of emissions.

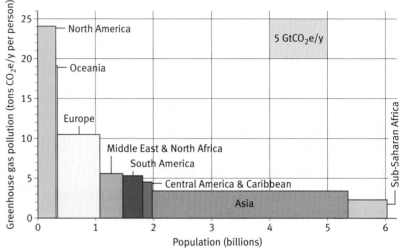

The UK is at risk too

The UK climate is kept mild by the Gulf Stream, a warm current from the Caribbean that flows towards Europe across the North Atlantic. There is evidence that this current slowed down in the past, making the UK an icy place. There are signs that the Gulf Stream may be slowing again.

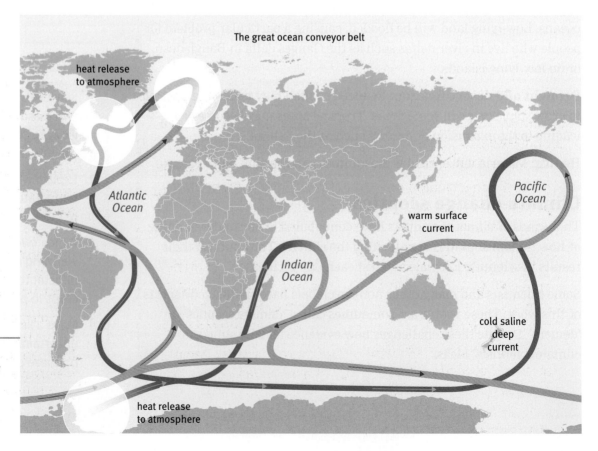

The Earth has a giant 'conveyor belt' system of ocean currents. It helps to warm land in northern latitudes.

What can governments do?

The UK government aims to reduce greenhouse gas emissions:

- 20% by the year 2020
- 80% by 2050.

The baseline is 592 million tonnes emitted in 1990. The graph shows there has been a small decrease in emissions.

It can be difficult for governments to change people's behaviour so that they produce less carbon dioxide. Democratic governments are sensitive to public opinion because they face election every few years. They may find it difficult to do what's best for the long term. People may protest and businesses may fight to protect their profits.

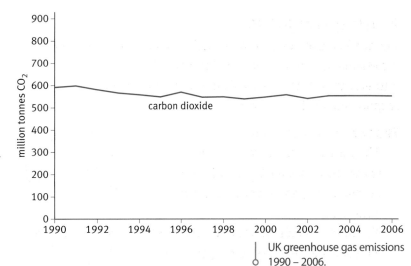

UK greenhouse gas emissions 1990 – 2006.

Technological solutions

There are many different proposals for using technology to reduce the amount of CO_2 entering the atmosphere. However, these may take a long time to develop and put into operation – and their outcomes may be uncertain.

Science to the rescue

Several solutions have been proposed that might get around the difficulties of reducing carbon emissions.

- Spread iron granules on the southern oceans. This would help the growth of plankton, which take dissolved CO_2 from the ocean. The oceans would remove more CO_2 from the atmosphere.

- Capture the CO_2 produced at power stations. Then compress it into a liquid and pump it into disused oil reservoirs beneath the sea-bed.

- Cement production counts for 5% of the greenhouse gases produced in Europe and America and more than 10% in China. A new type of 'eco-cement' absorbs CO_2 while setting and goes on absorbing CO_2 for years afterwards.

These are currently being tried and evaluated.

Extract from a popular science magazine.

Questions

5 Look at the four sources of emissions described in Section F . For each one, suggest what the government could do to reduce carbon emissions.

6 Look at the bar chart on the opposite page. What does the wide and low block for Asia show?

7 Do you think people should rely on technical solutions, like those suggested in the science magazine? Justify your answer.

We use aerials for sending and receiving radio signals to mobile phones. They are high up so that there is a good 'line of sight' between phone and aerial.

What is 'information'?

A mobile phone can store images (photographs), sounds (music and voice messages), text (messages), and numbers. Images, sounds, text, and numbers are all forms of **information**.

The phone can receive and transmit information, because it is part of a telephone network. It can also store and process information, for example, the user can amend a text message or play a game.

The phone is not full of pictures, sounds, and text. It stores the information electronically. This makes it easy to process the information, display it on the screen, or send it to another user.

Information paths

Mobile phones use microwaves. These are radio waves with the highest frequencies (and shortest wavelengths). Other radio waves are used to broadcast radio and TV programmes.

Radio waves and microwaves can travel for long distances through the air because they are only weakly absorbed by the atmosphere. This means that microwaves are also suitable for communicating with spacecraft far out in space.

Information is also transmitted along optical fibres, using visible light, infrared, or ultraviolet radiation. The fibres are made of high-purity glass, so that the radiation can travel many kilometres without being significantly absorbed.

Optical fibres have made possible cable television systems and the high-speed phone lines that are used by the internet.

Carrying information

Information is sent from place to place using a **carrier wave**. The radio waves, visible light or infrared can form the carrier wave. It must be modified to include the information. A wave carrying information is called a signal:

carrier wave + information = signal

You can send information at night using a flashing torch. The light from the torch is the carrier wave; the on-off flashes are the coded information. This is very similar to what happens in an optical fibre. A series of on-off pulses of light travel through the fibre and are received at the other end. An on-off signal, like the one below, is a **digital signal**. Just two symbols are needed to represent the signal: 0 (off, no pulse) and 1 (on, pulse).

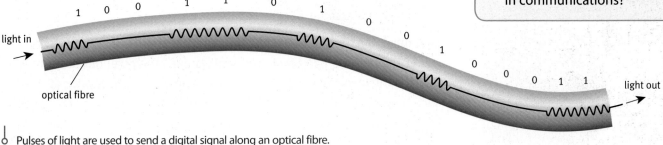

Pulses of light are used to send a digital signal along an optical fibre.

Coding information

All types of information can be converted into digital signals. A sound wave is an example. Speak into a microphone and your sound waves are converted into a varying voltage. The graph shows how the voltage might change during a fraction of a second.

The electrical signal has the same pattern as the original sound wave. A continuous variation like this is an **analogue signal**.

The table under the graph shows how this changing voltage can be converted into a digital signal. The voltage is sampled many times per second and the height of the graph (in volts) is coded as a binary number. These are the 0s and 1s that are used to switch the carrier wave on and off.

Images (such as photographs) can be coded by dividing them up into tiny dots (pixels). Each pixel is given a numerical value, which codes its colour and brightness.

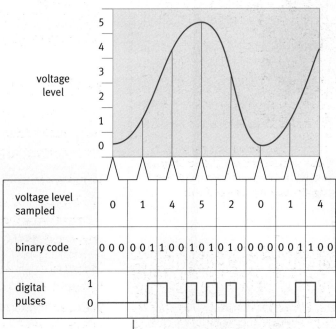

voltage level sampled	0	1	4	5	2	0	1	4
binary code	0 0 0	0 0 1	1 0 0	1 0 1	0 1 0	0 0 0	0 0 1	1 0 0
digital pulses								

How an analogue signal can be converted into digital pulses. (Real systems use more levels and a much faster sampling rate.)

aerial — | speaker

processor/
memory

1	2	3
4	5	6
7	8	9
*	0	#

keypad

microphone

The main components of a mobile phone.

Each pixel in the digital image has a number that gives information about the colour of that part of the picture. The bigger the choice of colours, the greater the range of numbers needed.

Radiation carries information

Are you receiving me?

If you send a message to a friend by flashing a torch, they must be able to decode it. Similarly, a mobile phone must be able to decode the digital signals it receives.

Inside the phone is a microprocessor that can do this. It simply reverses the process shown in the diagram on the previous page. It takes the binary codes and converts them into a varying voltage. This is sent to the earphone and you hear the sound.

More information

The amount of information needed to store an image or sound is measured in bytes (B). It takes about 1 megabyte (1 MB) to store one minute's worth of music. The memory of a hand-held mp3 player can vstore a hundred gigabytes of information. That is several weeks' worth of music, or thousands of photographs.

Each pixel in a digital image has information about the colour of that part of the picture. In a black-and-white image the pixel has a number to describe the brightness of the dot – from white through shades of grey to black. In a colour image the information also has to say what colour the dot is – this may be one of 256 colours or more than a million possible colours and shades. The colour image contains more information, which takes more time to send and more memory to store.

The pictures show two versions of the same picture.

Each pixel in the black-and-white image has one of 256 shades of grey between white and black; the information can be stored in 271 kB of memory. In the colour image each pixel can take over a million different colours and needs 74 MB of memory.

Mobile phones contain precious metals including gold, which can be recovered during recycling. Many people upgrade regularly as the technology improves.

The more information that is stored about an image or sound, the better quality it is.

Advantages of digital transmission

For transmitting information such as sounds and pictures, digital signals have several advantages over analogue ones.

- Digital signals can be processed by microprocessors (as in computers and phones).
- Digital information can be stored in memories that take up little space (as in computers, phones, and mp3 players).
- Digital signals can carry more information every second than analogue ones.
- Digital signals can be delivered with no loss of quality. Analogue signals lose quality, which cannot be restored.

Here is the reason for the last point. All signals get weaker as they travel along. **Noise** ('interference') also gets added in. A noisy analogue signal carrying music might sound blurry, scratchy, and distorted. But with digital signals, these effects can be corrected.

The diagram shows how. The noisy digital signal is passed through a regenerator. This is an electronic circuit that removes the noise. It can do this because it knows that the value of the signal must be either 0 or 1. A value close to 0 is corrected to 0, a value close to 1 is corrected to 1. It would take a very large amount of noise before it became difficult to tell the 0s and 1s apart!

Questions

4 Which contains more information, a 100 kB image file or a 1 MB sound file?

5 Explain why a 10 MB image file will produce a better picture than a 1 MB file of the same image.

6 Why is it possible to remove noise from digital signals but not from analogue ones?

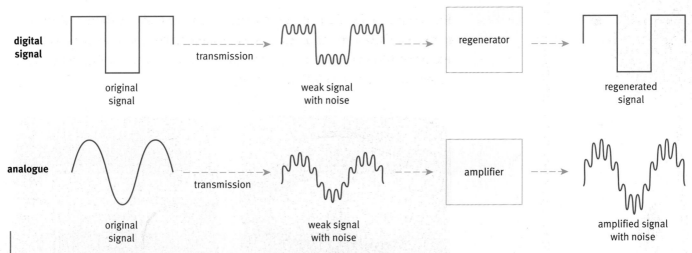

After transmission, a signal is weaker and noisier than the original. A digital signal can be 'cleaned up' by a regenerator. If an analogue signal is amplified to return it to its original height, the noise gets amplified as well.

Find out about

✔ **radiation from mobile phones and masts**
✔ **how to judge whether a health study is reliable**

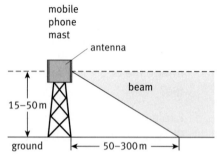

The microwave beam from a mobile phone mast.

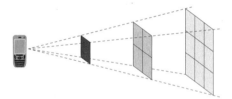

Twice as far away from the phone, the area is four times as great, so only a quarter of the radiation falls on each small square.

Some people think that road travel is less risky than going by train. But almost 3000 people die each year in UK roads. Far fewer people die on the railways. This is an example of the difference between a **perceived** risk and an **actual** risk.

Mobile phones are useful and they are fun. But are they dangerous?

A mobile phone sends out signals that are detected by a nearby mast. This is how the phone connects to the network.

Because the aerial is close to the user's ear, the user's head will absorb some of the energy of the microwave radiation. The amount of energy absorbed depends on the intensity of the radiation and the length of time of the exposure.

Absorbed radiation has a heating effect so, as we saw on page 55, the user's head is slightly heated by the radiation it absorbs. This depends on the number of photons absorbed and the energy each delivers.

Radiation spreading out

The intensity of the radiation coming from a mobile phone is greatest as it leaves the phone. Because the radiation spreads out in all directions, it rapidly gets weaker the further away it travels from the phone.

Similarly, the radiation from the mast is most intense as it leaves the mast. It is more intense than the radiation from a phone. By the time it reaches a distant phone, it is very weak.

People have concerns about radiation from phone masts. If you stood right next to the mast, the heating effect of the radiation absorbed by your body could be quite noticeable.

Fortunately, you cannot get that close. Phone masts are designed so that their radiation is shaped like the beam of light from a lighthouse. If you stand directly under a mast, its radiation is much weaker than the radiation from your phone.

People may not accurately estimate a particular risk. People tend to overestimate the risks of things with invisible effects, like radiation. Also, they overestimate the risks of things where they feel less in control. For example, people worry more about flying than cycling, although many more people are killed or injured in cycling accidents.

Health studies

Over 50 million people in the UK use mobile phones. Few worry about any unknown risks. People like the benefits a mobile phone brings. Research has so far failed to show that there are any harmful effects.

To look for any harmful effects, scientists compare a sample of mobile phone users with a sample of non-users. Does one group show a higher rate of cancer, for example?

Can we have confidence in the results?

The news often has reports of studies that compare samples from two groups, to see if a particular factor or treatment makes a difference. To judge whether **we can have confidence in the results** of studies like this, there are two things worth checking.

What to check ...	... and why
Look at how the two samples were selected. Can you really be sure that any differences in outcomes are really due to the factor claimed?	A study to find whether mobile phone use caused cancer would need to compare samples of users and non-users.
	The samples should match as many other factors as possible, for example, similar numbers of young people in each sample. This is because the development of brain tumours might be age-related.
	Samples should be selected randomly, so that other factors, such as genetic variability, are similar in both groups.
Were the numbers in each sample large enough to give confidence in the results?	With small samples, the results can be more easily affected by chance. Larger sample sizes give a truer picture of the whole population. The effect of chance is more likely to average itself out.

How great is the risk?

Health outcomes are often reported as relative risks. For example, 'people exposed to high levels of sunlight were four times as likely to develop eye cataracts'.

- If your risk was one in a million, it rises to four in a million – not a worry!
- If your risk was 5 in 100, it rises to 20 in 100 – worth avoiding!

And some people might be more at risk than others, for reasons of family history or lifestyle.

Key words
- ✔ reliable
- ✔ perceived risk
- ✔ actual risk

Questions

1 Is there a health risk from low-intensity microwave radiation from mobile phone handsets and masts? (The answer is neither 'Yes' nor 'No'.)

2 a Look at the first row of the table. What outcome and what factor that might cause it are being studied?
  b Describe a second way that the samples should be matched in this study.
  c Explain why it is important to match the two samples.

Science Explanations

In this module you will learn about the electromagnetic spectrum, different types and sources of radiation, and how it can be both useful and dangerous.

You should know:

- how to think about any form of radiation in terms of its source, its journey path, and what happens when it is absorbed, transmitted, or reflected
- that a beam of electromagnetic radiation delivers energy in 'packets' called photons
- the parts of the electromagnetic spectrum, in order of their photon energies
- what different parts of the electromagnetic spectrum can be used for
- two factors that affect the energy deposited by a beam of electromagnetic radiation
- how the intensity of an electromagnetic beam changes with distance
- that the three parts of the electromagnetic spectrum with highest photon energies are ionising
- why ionising radiation is hazardous to living things
- how people can be protected from ionising radiation
- how microwaves heat materials, including living cells
- about the features of microwave ovens that protect users
- that sunlight provides the energy for photosynthesis and warms the Earth's surface
- how photosynthesis affects which molecules are in the atmosphere
- what the greenhouse effect is (and be able to identify greenhouse gases)
- how to use the carbon cycle to explain how green plants and decomposers have kept the carbon dioxide concentration constant for thousands of years
- how the atmosphere's ozone layer protects living organisms from ultraviolet radiation
- about global warming and its possible effects on agriculture, weather, and sea levels
- that computer models provide evidence that human activity causes global warming
- that radio waves and microwaves carry information for radio and TV through the atmosphere and through space
- that infrared and visible light waves carry information along optical fibres
- that a sound wave can be superimposed onto an electromagnetic wave, and that this signal can be carried, and then decoded to produce a copy of the original sound
- that this coding can be an analogue signal, which varies continuously, or a digital signal, which is a series of pulses
- the advantages of digital signals in reducing noise, ease of storage, and manipulation of the stored signals.

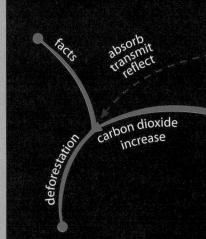

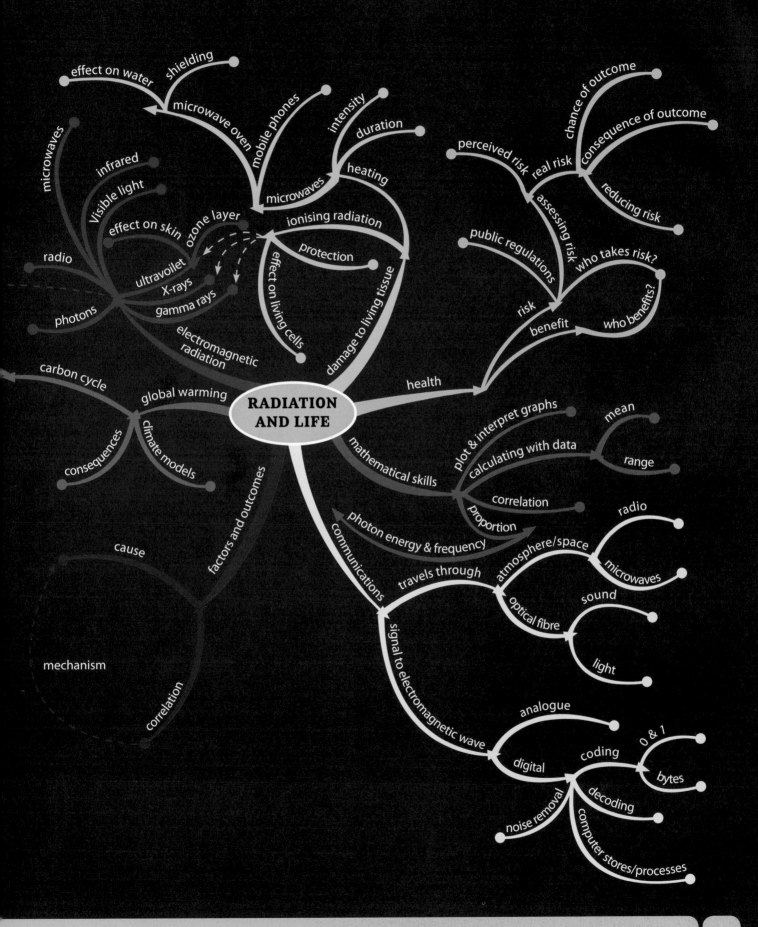

Ideas about Science

Besides developing an understanding of the electromagnetic spectrum, it is important to recognise the difference between correlation and cause, and to assess the risks and benefits associated with the electromagnetic spectrum.

Factors and outcomes may be linked in different ways, and it is important to distinguish between a correlation, where a change in one is linked to a change in the other, from a cause, where there is a method by which the factor is responsible for the outcome.

In the context of the electromagnetic spectrum, you should be able to:

- suggest and explain everyday examples of correlation
- identify a correlation from data, from a graph, or from a description
- suggest factors that might increase the chance of a particular outcome, but not invariably lead to it
- identify that where there is a machanism to explain a correlation, scientists are likely to accept that the factor causes the outcome.

Everything we do carries some risk, and new technologies often introduce new risks. It is important to assess the chance of a particular outcome happening, and the consequences if it did, because people often perceive a risk as being different from the actual risk: sometimes less, and sometimes more. A particular situation that introduces risk will often also introduce benefits, which must be weighed up against that risk.

You should be able to:

- identify risks arising from scientific or technological advances
- interpret and assess risk presented in different ways
- discuss risk, taking into account both the chance of it occurring and the consequence if it did

- discuss both the risks and benefits of a course of action, taking into account of who takes the risk and who benefits
- distinguish between real risk and perceived risk
- suggest why people are willing (or reluctant) to take certain risks
- discuss how risk should be regulated by governments and other public bodies.

Review Questions

1 **a** Copy out and complete the diagram of the electromagnetic spectrum. Use words from this list.

microwaves   **gamma rays**   **ultraviolet**

radio waves		infrared	visible light		X-rays	

b Where in the electromagnetic spectrum do the photons have the most energy?

c Which **one** of the following is **not** ionising radiation?

gamma rays   **microwaves**   **ultraviolet**   **X-rays**

2 The graph shows how the percentage of carbon dioxide in the atmosphere has changed over the past 300 years.

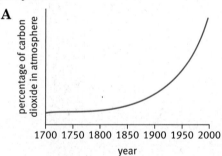

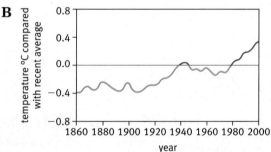

a Describe the trend in graph A and explain the scientific reasons for what the graph shows.

b Graph B shows how the average temperature of the atmosphere has changed over the past 140 years. Use the two graphs to explain the meaning of **correlation**.

3 Match the wave diagrams to the statements opposite. Diagrams may be used once, more than once, or not at all.

A    D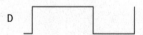

B    E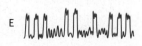

C    F

a A sound wave is an analogue wave.

b The sound wave is converted into a digital code. The digital signal is sent as a series of short pulses.

c Digital signals can be transmitted with higher quality than analogue signals. As the signal is transmitted, it decreases in intensity and picks up noise.

d When the signal is received it is amplified.

e The signal is cleaned up to remove the noise.

f The digital signal is then decoded to reproduce the original sound wave.

P3 Sustainable energy

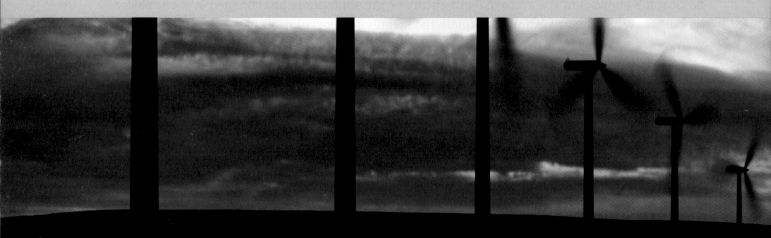

Why study sustainable energy?

Energy supply is one of the major issues that society must address in the immediate future. To make decisions about energy use, you need to understand the figures and calculations behind the headlines. Electricity supplies many of our energy needs. Most of us take electricity for granted. But today's power stations are becoming old and will soon need replacing. How should we generate electricity in the future? Can we reduce our impact on the environment without reducing our quality of life?

What you already know

- When energy is transferred the total amount of energy remains constant.

- Whenever energy is transferred some of it spreads out into the surroundings.

- Electricity is a useful way of transferring energy long distances.

- Electricity is generated in power stations using a variety of energy sources.

- Electric current transfers energy from the power supply to devices in the circuit.

- The higher the voltage of the power supply the more energy the current transfers.

Find out about

- how much energy we use, as individuals, as a country, and across the world

- how we could use energy more efficiently

- how electricity is generated in a power station

- the choices for generating electricity in the future.

The Science

Energy cannot be created or destroyed, but whenever we use it some is lost to the surroundings. It is difficult to recover to use again. We need to find ways of using energy more efficiently. Most UK electricity is generated by burning gas and coal to drive generators. To choose the right generation methods for the future you need to know about the advantages and drawbacks of each.

Ideas about Science

Nothing can be completely safe; there are different risks associated with using each energy source. But who should be making the decisions? How can you have your say?

Find out about

- ✔ **energy sources**
- ✔ **why we need to be concerned about energy supplies**

Key words

- ✔ **conserved**
- ✔ **energy source**
- ✔ **primary source**
- ✔ **secondary source**
- ✔ **fossil fuel**
- ✔ **biofuel**
- ✔ **pollutant**
- ✔ **sustainable**

People need energy to keep them alive, warm, and able to move. Over time, demand for energy has grown. There are more people now than at any time in the past, and the population continues to expand. Modern transport, buildings, possessions, and communications need more energy than ever before. People travel further and faster and have different expectations about their lifestyle. Understanding about the energy sources available is important when making choices.

Modern living can demand large amounts of energy.

Energy sources

Energy is **conserved**, meaning it can neither be created nor destroyed. The energy we use for heat, movement, and light must all come from an **energy source**. For example, we can release energy by burning fuels, or we can use energy carried by radiation from the Sun.

A **primary energy source** is one that is found or occurs naturally. Examples include fuels such as coal, oil, natural gas, and wood. Wind, waves, and sunlight are also primary energy sources.

Nowadays, much of our energy use involves electricity. Electricity is a **secondary energy source**. It must be generated using a primary source.

Nations and individual people pay for the energy they use. The price is related to the amount of fuel burned, and to the cost of distributing energy to the users. Energy bills are a major expense for most households and for the country as a whole.

Most UK electricity is generated by burning fossil fuels.

Fuels

A **fossil fuel** is one that has built up over millions of years by the decay of plant and animal remains. Coal, petroleum, and natural gas are all fossil fuels. We are using them up far more rapidly than they can form. The table shows how long the world's petroleum supplies will last.

Country	Number of years' supply left (from 2010)
Saudi Arabia	70
Canada	147
Iran	93
Iraq	148
Kuwait	108
United Arab Emirates	91
Venezuela	86
Russia	15
Libya	64
Nigeria	39

Most of the world's petroleum is found in these ten countries.

When they burn, fossil fuels produce carbon dioxide (CO_2) and other **pollutants**, such as carbon particles. The amount of CO_2 in the atmosphere has risen over the past two hundred years and is affecting the Earth's climate.

A **biofuel** is one that has recently come from living material (biomass). Wood, straw, sewage, and sugar are all used as biofuels. Like fossil fuels, biofuels produce CO_2 when they burn. Unlike fossil fuels, biofuels are produced quickly.

Nuclear fuel releases energy without burning so it does not make CO_2. Nuclear fuels are not found in the UK so any that we use must be imported.

Sustainable energy

Our current use of energy sources is not **sustainable** and cannot continue indefinitely. Some sources are running out, and some of our energy use is damaging the environment. Governments and individuals must decide how to reduce demand for energy and plan which energy sources to use in the future. To decide what to do, we need to know some facts about the amount of energy we use.

Petroleum is an oily mixture of solid, liquid, and gas. Petrol, oil, and diesel fuel are all made from petroleum.

Can wind power supply enough energy?

Questions

1 Write down three things that you do during a day that use:
   a a primary energy source
   b a secondary energy source.

2 Suggest at least two reasons for reducing our use of fossil fuels.

Find out about

- ✔ **what is measured on a domestic electricity meter**
- ✔ **how to calculate the energy used by an electrical appliance**
- ✔ **how to calculate the cost of the electrical energy used**
- ✔ **how electric power is related to current and voltage**

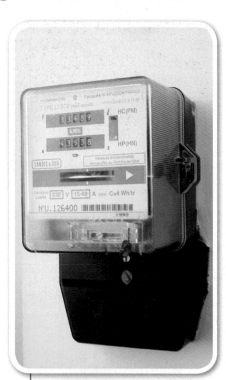

The electricity meter in your home measures the number of kilowatt-hours of electrical energy that you buy.

Meter: 326565		Tariff: **Domestic**	
cost of energy	**number of units used**	**unit charges**	**total**
13.25	2213	first 227 at 13.25p	£30.08
		next 661 at 7.88p	£52.08
			£82.16

We can calculate how much energy appliances use and how much they cost to run. This information can help us make sense of our electricity bills.

Measuring energy

The energy used by an electrical appliance can be measured with a meter. This energy depends on:

- the power rating of the appliance
- the time it is on for.

Power is the rate at which energy is used. An appliance with a rating of 1 **watt** (1 W) uses 1 **joule** of energy (1 J) every second. But 1 watt is a very small power, 1 second is a very short time, and 1 joule is a tiny amount of energy.

Many domestic appliances have powers of a few kilowatts. One kilowatt is one thousand watts. For domestic appliances it is more convenient to use the **kilowatt-hour** as the unit of energy. It is the energy used by a 1 kW appliance switched on for one hour.

$$1 \text{ kWh} = 3\,600\,000 \text{ J}$$

energy used	=	power rating	×	time
(joule, J)		(watt, W)		(second, s)
(kilowatt-hour, kWh)		(kilowatt, kW)		(hour, h)

Paying for energy

On an electricity bill, '1 unit' means 1 kilowatt-hour. To find the cost of the energy, multiply the number of units by the cost of one unit.

cost of energy = number of units used × price of one unit

(pence, p)   (kilowatt-hours, kWh)   (pence per kilowatt-hour, p per kWh)

Worked example

Working out the cost

A 3 kW immersion heater heats water for a bath in 2 hours. Electricity costs 11p per unit.

Energy used = 3 kW × 2 h = 6 kWh

Cost = 6 kWh × 11 p/kWh

= 66 p

Power, current, and voltage

When an electrical appliance is switched on, electric **current** passes through it and energy is transferred to the appliance and its surroundings. The power is related to the current and to the **voltage** of the electricity supply.

power = voltage × current
(watt, W)   (volt, V)   (amp, A)

In the UK, the mains electricity voltage is 230V. For a battery-operated device, the voltage is usually just a few volts.

Appliances that include motors, or are used for heating, usually need quite large currents. They have high power ratings and are expensive to run.

Every electrical appliance has a power rating in watts or kilowatts. This tells you how much energy it uses each second when it is switched on.

Worked example

Working out the current

A hair dryer has a power rating of 700 W.

power = voltage × current

For all UK mains appliances the operating voltage is 230 V.

700 W = 230 V × current

Divide both sides by 230 V to find the current.

$$current = \frac{700\ W}{230\ V}$$

current = 3.0 A

Key words

- ✓ **power**
- ✓ **watt**
- ✓ **joule**
- ✓ **kilowatt-hour**
- ✓ **current**
- ✓ **voltage**

Questions

1 Look carefully at each of the tasks in the table on the right that use electricity, and put them in the order you think they would come – from the cheapest to the most expensive. Then calculate the number of kilowatt-hours that each involves and see if your estimate of the cost was correct.

2 Electricity costs 10p per unit. Calculate how much it would cost to use a fan heater for 2 hours.

3 What is the power of a mains appliance that needs a current of 3 A to make it run?

Task	Appliance used	Power rating (W)	Time for which it is on
watch television for the evening	television	300	5 hours
dry your hair	hairdryer	700	5 minutes
make a pot of tea	electric kettle	2000	4 minutes
write a homework assignment	computer	250	2 hours
listen to music	mp3 player	0.2	2 hours
heat your bedroom while you do your homework	electric fan heater	1500	2 hours
wash a load of dirty clothes	washing machine	1850	$1\frac{1}{2}$ hours
play a game	games console	190	1 hour

We use far more energy in a day than is accounted for by the electricity bill at the end of the month. How much energy do you use in a day?

Heating and cooking

Most of the energy you use at home is probably supplied by electricity. For some tasks you might use another source such as gas or oil, but the energy needed will be the same.

Task	Energy (kWh)
Bath (about 100 litres of hot water)	5
Shower (about 30 litres of hot water)	1.4
Gas cooker (for 1 hour)	1.5
Room heater (eg radiator) (for 1 hour)	1
Patio heater (for 1 hour)	15

Transport

Different means of transport use different amounts of energy.

Burning 1 litre of petrol in a car releases about 10 kWh of energy. An economical car can travel about 10 miles (16 km) on 1 litre of fuel, so each mile needs about 1 kWh, which is about 0.6 kWh per km.

Travelling alone in a car uses much more energy than sharing public transport. So figures for public transport are based on having a full vehicle. The table lists energy per **passenger-kilometre**, which is each passenger's share of the energy used to travel 1 km.

A large, less economical car needs about 1.3 kWh per mile (about 0.8 kWh per km).

Food and drink

Growing and producing food uses energy. Some of that energy is stored in the food and passes on to you when you eat it. The table lists the energy needed to produce some fresh foods. Processed foods use more energy.

Transport	Energy per passenger-km (kWh / passenger-km)
Bus	0.19
Train	0.06
Aircraft	0.51
Boat	0.57

Food	Energy (kWh) for production
1 egg	0.5
1 pint of milk	0.8
50 g cheese	0.8
100 g meat (eg beef, chicken, pork)	4
100 g fruit or vegetables	0.5

Other stuff

Everything you use has an energy cost. The table lists the energy needed to make and transport some items that you might buy or use.

Item	Energy (kWh)
Drinks can	0.6
Plastic bottle	0.7
AA battery	1.4
Magazine	1.0
Computer	1800

To find the daily energy cost of an item that you keep for more than a day:

$$\text{Daily energy use (kWh per day)} = \frac{\text{energy needed to produce the item (kWh)}}{\text{number of days that the item lasts}}$$

An average person in the UK uses about 160 litres of clean water per day. The energy needed to treat and distribute this amount of water is the **energy cost** of the water. In the UK the energy cost of our daily water use is about 0.4 kWh.

Worked example

Calculating the energy used

Two people travel 5 km to school in the car.

energy used (kWh) = distance travelled (miles) × energy per km (kWh per km)

energy used = 5 miles × 0.6 kWh per km = **3 kWh**

If they had travelled by bus then

energy used (kWh) = distance travelled (km) × number of passengers
× energy per passenger-km (kWh per passenger-km)

energy used = 5 km × 2 passengers × 0.19 kWh per passenger-km

energy used = **1.9 kWh**

On average each person in the UK throws away 400 g of packaging per day with an energy cost of about 4 kWh.

Questions

1 An aircraft flies from London to New York and back, a total of 5586 km. Look at the information about energy used for transport.
   a If the aircraft carries 500 people what is the total energy used?
   b Explain why the energy cost would be more than half your answer to part a if there were only 250 passengers.

2 Suggest reasons why producing processed food needs more energy than fresh food.

3 Someone wants to change their lifestyle and use less energy. Use information from these pages to suggest what they might do.

Find out about

- ✔ **how public services and activities contribute to our energy use**
- ✔ **how people's daily energy use varies between countries**

In the UK, the average daily energy use per person is about 110 kWh per person per day. A diary of your own energy use will probably give a much lower figure. Where is the rest of the energy used? Are we using more than our 'fair share'?

The bigger picture

We can calculate the average energy used per person in the UK by sharing the total between everybody in the country.

The armed forces work on behalf of everyone in the country. Our share of their energy use is about 4 kWh per person per day.

Building homes uses about 1 kWh per person per day.

Building and maintaining roads uses about 2 kWh per person per day.

Computer servers are at the core of many businesses and at the heart of the internet. They need energy to drive the computers and even more energy to cool them. Servers across the UK use about 0.5 kWh per person per day.

Supermarkets use about 0.5 kWh per person per day.

Global issues

The map shows the average energy use per person in different parts of the world.

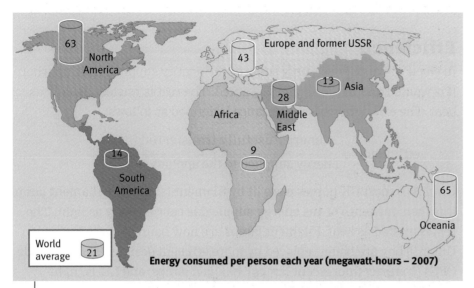

As countries become industrialised, living standards rise and energy use increases. Energy use in India and China is now growing especially fast.

The table shows the average daily energy use for people in various countries. The figures for gross domestic product (GDP) per head indicate how rich a country is.

In general, richer countries use more energy per person than poorer ones. If people have more money they can buy more goods, live in larger, more comfortable houses, and travel more. All these things use energy.

But the figures for daily energy use still do not tell the whole story, because they only include energy used within each country. They do not include energy used to make imported goods. To take account of all the UK imports, we should add about 40 kWh per person per day to the UK energy figure.

Country	Average daily energy use (kWh per person per day)	GDP per person ($)
Australia	190	47 000
China	50	3 000
Denmark	120	62 000
France	140	45 000
India	20	1 000
Japan	130	38 000
Kuwait	300	54 000
Mexico	60	10 000
Poland	80	14 000
Turkey	40	10 000
UK	110	44 000
USA	250	48 000

Energy use per person and GDP per person 2007 source: World Bank.

These jeans are made in China. The energy used to grow the cotton, weave the cloth, and make the garment contributes to the average energy use in China, not the UK.

Questions

1   Suggest at least two more energy-using activities that are shared between everyone in the UK.

2   Plot a scatter graph of daily energy use against GDP for the countries listed in the table. Comment on any trend shown by your graph.

3   'Our energy use is part of a global problem. We should be part of the solution.' Do you agree? Write a paragraph giving reasons for your views.

Find out about

- ✔ what is meant by 'efficiency'
- ✔ how to use a Sankey diagram to show energy transfer

We can use less energy by switching off appliances. But we should also use appliances that don't waste energy – appliances that are more efficient.

Efficiency

In electrical appliances, only some of the energy supplied ends up where it is wanted and in the form it is wanted. The rest is wasted, usually as heat. The **efficiency** of an appliance is defined as follows:

$$\text{efficiency} = \frac{\text{energy usefully transferred}}{\text{energy supplied to the appliance}} \times 100\%$$

Until 2009 most UK homes were lit by filament lamps. In a filament lamp, less than one tenth of the energy supplied is carried away as light. The rest is wasted as heat. Filament lamps are now banned in the European Union. They are being replaced by more efficient designs. These include CFLs (compact fluorescent lamps), halogen lamps, and LEDs (light-emitting diodes).

The CFL on the left needs less energy than the filament lamp on the right to produce the same light output per second. It is more efficient.

Worked example

Calculating efficiency

An 600 W electric motor is used to lift a load. In one minute the load gains 18 000 J of gravitational potential energy. How efficient is the motor?

Calculate the energy supplied: energy = power × time

$$\text{energy} = 600 \text{ W} \times 60 \text{ s} = 36\,000 \text{ J}$$

Calculate the efficiency:

$$\text{efficiency} = \frac{\text{energy usefully transferred}}{\text{energy supplied to the motor}} \times 100\%$$

$$\text{efficiency} = \frac{18\,000 \text{ J}}{36\,000 \text{ J}} \times 100\%$$

$$\text{efficiency} = 50 \%$$

Question

1 A CFL rated 20 W gives a light output of 11 W.
   a How much energy is used each second by the lamp?
   b How much of this energy is useful?
   c Calculate the efficiency of the lamp.

Sankey diagrams

In a **Sankey diagram**, branching arrows show how energy is transferred. Their width indicates the amount of energy. The total width stays the same because energy cannot be lost or gained overall.

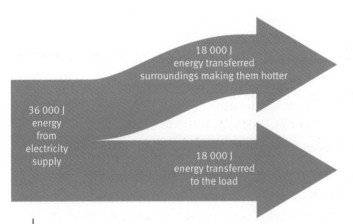

The Sankey diagram for the motor lifting a load with efficiency of 50%.

Electrical appliances are labelled with an efficiency rating to help people choose which to buy.

Questions

2 A filament light bulb rated at 100 W gives a similar output to the 20 W CFL lamp in question 1.
   a Suggest why filament lamps should no longer be sold.
   b Draw Sankey diagrams for both lamps to illustrate your answer.

3 An electric kettle rated 3 kW takes 3 minutes to heat some water. 50 000 J of energy is transferred to the surroundings.
   a How much energy does the electricity supply deliver to the kettle?
   b How much useful energy is transferred to the water?
   c What is the efficiency of the kettle?
   d Draw a labelled Sankey diagram for the kettle.

4 A fundamental law of physics is that energy is always conserved. Energy cannot be created or destroyed. Explain how a Sankey diagram shows this.

5 Explain how using energy-efficient appliances is an advantage to the person and also to the country.

Key words
✔ **efficiency**
✔ **Sankey diagram**

Find out about

- ✔ the main types of fuel used in the UK
- ✔ the advantages of using electricity
- ✔ how a magnet moving near a coil can generate an electric current
- ✔ how this is used to generate electricity on a large scale

What fuel do we use and where?

Each year the UK government publishes information about the country's energy use. They refer to three main energy sources: electricity, natural gas, and petroleum (petroleum includes petrol, diesel, and all other fuels made from oil). Other fuels, including solid fuels such as coal and wood, are all grouped together as 'other'.

The main UK energy users are industry, transport, and domestic. Each has a different pattern of fuel use.

Energy use by UK industry 2008.			
Electricity	Gas	Petroleum	Other
33%	38%	21%	8%

Energy use by UK transport 2008.			
Electricity	Gas	Petroleum	Other
1%	0%	97%	2%

UK domestic energy use 2008.			
Electricity	Gas	Petroleum	Other
22%	69%	7%	2%

Questions

1 Write down four tasks that you normally do, using electricity. For each one say whether you could easily use a primary fuel instead.

2 Draw pie charts to show energy use by UK industry in 2008. Draw similar charts for transport and for domestic use.

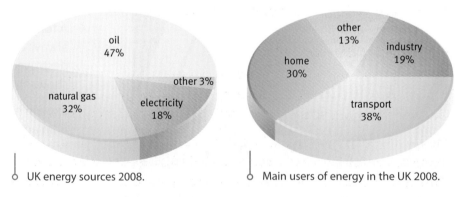

UK energy sources 2008.

Main users of energy in the UK 2008.

In our homes, we might use gas, oil, or other fuels for heating and cooking. For almost everything else we use a secondary source – electricity. Electricity can be used for many different tasks and it is easy to distribute using cables and wires.

Generating electricity

Nowadays most of us in Britain take a mains electricity supply for granted. But it was only in May 2003 that Cym Brefi in mid-Wales became the last village in Britain to get a mains electricity supply.

Electricity reaches last village in Britain

Not having a mains electricity supply does not mean you cannot use electrical appliances. Many can be run from batteries. But this works only for relatively low-power devices. For others, you might use a diesel-powered **generator**.

This is how the inhabitants of Cym Brefi ran their washing machines and vacuum cleaners before they got mains electricity. But generators are noisy, and each 'unit of electricity' is much more expensive than from the mains. So they can only be run for a short time.

A simple generator

Generators work on the principle of **electromagnetic induction**. This phenomenon, which does so much to make our lives comfortable and convenient, was discovered in the 1830s by Michael Faraday.

One way to generate a current is to move a magnet into, or out of, a coil. The movement of the magnet causes an induced voltage across the ends of the coil. 'Induced' means that it is caused by something else – in this case, the movement of the magnet. The coil, for a brief time, is like a small battery. If the coil is part of a complete circuit, this induced voltage makes a current flow.

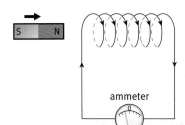

ammeter

1 While the bar magnet is moving into the coil, there is a small reading on the sensitive ammeter.

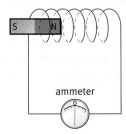

ammeter

2 There is no current while the magnet is stationary inside the coil.

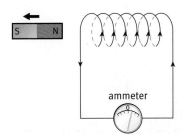

ammeter

3 While the magnet is being removed from the coil, there is again a small current, but now in the opposite direction.

Moving a magnet into, or out of, a coil generates a current.

Key words

- ✓ generator
- ✓ electromagnetic induction
- ✓ alternating current (a.c)

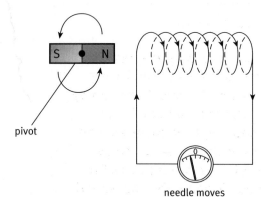

rotate magnet

pivot

needle moves
back and forth

A rotating magnet generates an alternating current in the coil.

Continuous current

If a magnet is repeatedly moved in and out of a coil, or rotated close to a coil, a continuous to-and-fro current (an **alternating current, a.c.**) can be generated. This is what happens inside a shake torch, a wind-up radio, some bicycle light systems, and in large-scale electricity generators.

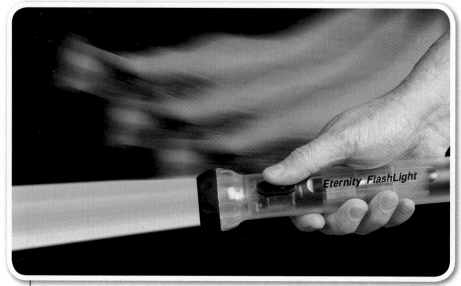

Shaking this torch moves a magnet in and out of a coil to generate an electric current.

front light

rear light

moving magnet

generator

This bicycle has a small generator that uses the movement of the wheel to produce a current. The generator is connected to a rechargeable battery that supplies its lights.

The technician is constructing this generator. Wires are wound around the outside. A turbine will make magnets rotate in the centre, inducing an electric current in the wires.

Questions

3 In a shake torch, how will the current change as the torch is shaken more vigorously? Explain why.

4 Suggest a situation where a wind-up radio would be more convenient than an ordinary radio with batteries.

5 Explain why the bicycle light system shown in the picture needs a rechargeable battery.

Human power

Can we be self-sufficient in energy? Instead of using mains electricity from power stations, could we power all our appliances ourselves?

Pedal power station

In 2009, the BBC television programme 'Bang Goes the Theory' set up a human power station. To generate electricity, 70 cyclists pedalled bicycles. The electricity was used to power appliances in a family house.

As high-power appliances were switched on, the cyclists found it harder to pedal enough to supply the power. The oven and the power shower were the most difficult appliances to run with cycle power. The greater the current supplied by the generator, the harder the cyclists had to pedal.

The cyclists became tired, hot, and sweaty. They needed to eat and drink in order to get enough energy to keep pedalling.

Generating electricity is never 100% efficient. Only some of the energy stored in the cyclists' bodies from the food they had eaten was used to produce electricity. Quite a lot was carried away as heat.

Questions

6 Sketch a Sankey diagram for a cyclist in the human power station. The input is the energy stored in their body from the food they have eaten, and the useful output is the energy carried by the electricity. On your diagram label the wasted energy.

7 A fit cyclist can produce an output power of 200 W.
  a If they can keep this up for 24 hours non-stop, what is their energy output in kWh?
  b If you need 125 kWh per day, how many cyclists would you need to be your 'slaves'?

Each bicycle in the human power station was fixed in place with the back wheel connected to a small generator.

Who decides?

Electricity is a secondary energy source. Energy companies, operating under government regulation, generate and distribute it.

Energy companies also make decisions on your behalf. When you boil a kettle, the electricity may have come from any type of primary source.

Burning fuel

In a fossil-fuel power station, coal, gas, or oil is burned to boil water and make high-pressure steam. Biofuels, such as wood, can be used in the same way. In some places, heat is extracted from underground rocks to heat water; this is geothermal energy. Any power station that works like this is known as a **thermal power station**.

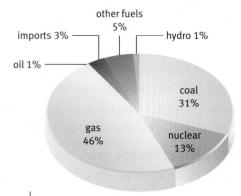

other fuels 5%

imports 3%

hydro 1%

oil 1%

coal 31%

gas 46%

nuclear 13%

In the UK, most electricity nowadays is generated by burning fossil fuels.

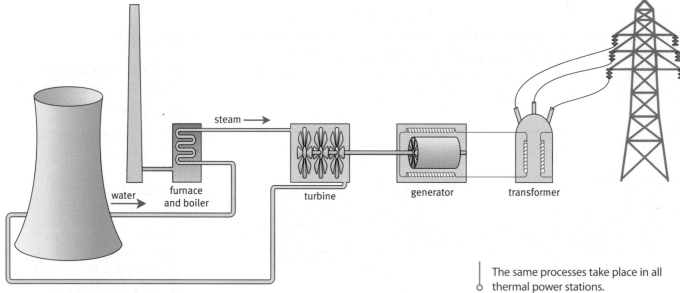

water

furnace and boiler

steam →

turbine

generator

transformer

The same processes take place in all thermal power stations.

Thermal power station

In a thermal power station, high-pressure steam passes through a **turbine**. The turbine rotates a generator to produce electricity. In UK power stations the rate of turning is set at 50 cycles per second.

After passing through the turbines, the steam condenses to water. It can be fed back into the boiler and used again.

This turbine has many small blades that are driven around by the steam.

Regular maintenance keeps the generators running smoothly.

Steam collects in cooling towers where it condenses back to water.

Reducing waste

In any fossil-fuel or biofuel power station, only some of the energy from the burning fuel is transferred electrically. A lot of energy is wasted because it is carried away as heat in steam and exhaust gases.

Burning fuels produce CO_2 and other waste products. Some of the worst pollutants are removed from the exhaust gases before they can escape into the atmosphere.

Using more efficient power stations is one way of reducing the amount of CO_2 produced. Power stations burning natural gas have an extra turbine that is driven by the hot exhaust gases. This makes them the most efficient type of fossil fuel power station. But there are arguments both for and against building more gas-fired power stations.

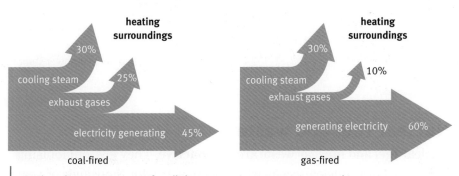

Sankey diagrams account for all the energy. Less energy is wasted in a gas-fired power station.

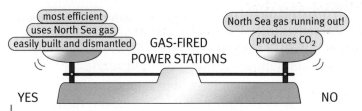

Weighing the arguments – should we build more gas-fired power stations?

Key words

✓ **thermal power station**
✓ **turbine**

Questions

1 What percentage of UK electricity is generated using fossil fuels?

2 According to the Sankey diagrams on this page, what is the typical efficiency of a coal-fired power station?

3 Sketch a Sankey diagram to show where the energy goes in a typical oil-fired power station with an efficiency of 38%.

4 What might be the benefits and drawbacks of using biomass fuel in power stations? Draw a balance diagram to summarise your ideas.

There are ten nuclear power stations operating in the UK, some of these are coming towards the end of their working lives. In 2009 the government proposed that another ten should be built and in operation by 2020. They said that these would be essential to meet CO_2 emission targets. Is more nuclear power the right choice?

Nuclear power stations

Nuclear power stations use solid fuel that contains uranium. In a **nuclear reactor**, uranium atoms split into lighter atoms, releasing energy, so the fuel becomes very hot. The hot fuel boils water to make steam that drives turbines.

As the uranium atoms split, the fuel gradually becomes solid nuclear waste.

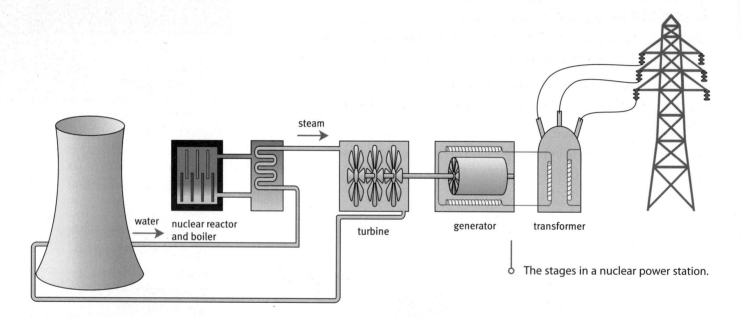

steam

water | nuclear reactor and boiler

turbine

generator

transformer

○ The stages in a nuclear power station.

○ Hot waste from nuclear fuels is stored under water.

Nuclear fuel and waste

Nuclear fuel and nuclear waste are **radioactive**. They give out ionising radiation. Some nuclear waste will be radioactive for thousands of years. Waste from nuclear fuel is very hot at first, so it is stored under water until it cools. When cool, waste is mixed with concrete and stored to make sure it does not contaminate the atmosphere, water supply, or soil.

Nuclear sites are monitored to make sure that workers and the public are not put in danger from radioactive material. Radioactive **contamination** occurs when radioactive material lands on or gets inside something. Exposure to ionising radiation is called **irradiation**. Limits for irradiation are set by law.

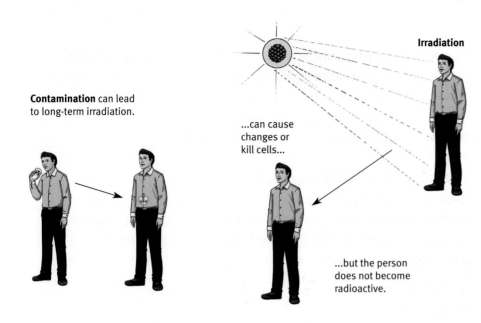

Contamination can lead to long-term irradiation.

Irradiation

...can cause changes or kill cells...

...but the person does not become radioactive.

Uranium mines contain enough fuel for hundreds of years.

Benefits and risks

Nuclear fuel yields far more energy than the same amount of fossil fuel. 1 g of uranium fuel can provide as much energy as 8 kg of fossil fuel.

A nuclear power station produces much less waste than a fossil fuel power station. A nuclear reactor does not burn fuel, so no CO_2 is produced.

Suppose there is an accident and fuel leaks out?

We can't see radiation, how can we judge the risk?

There are no uranium mines in Britain. I am worried about relying on imports.

Why should we be exposed to risks so that everyone can have cheap electricity?

What if nuclear waste falls into the hands of terrorists?

They can't store waste at Sellafield, with sea levels rising. It's on the coast!

Many people are concerned about nuclear power stations.

Questions

1 Why might drinking water, contaminated by radioactive waste, be more dangerous than being irradiated by the glass of water?

2 Do you think that the UK government should build more nuclear power stations? Write an entry for a blog to persuade other people to agree with you.

Find out about

- ✓ **how renewable energy sources work**
- ✓ **how much UK electricity is generated from renewable sources**

A **renewable** energy source is one that can be used without running out. We already use some renewable energy sources in the UK. Should we use more?

Solar power

In the UK, electromagnetic radiation from the Sun provides an average of about 100 W of **solar power** per square metre of ground. Solar **thermal** panels use the Sun's radiation to heat water or buildings directly. Covering *all* south-facing roofs in the UK with thermal solar panels could provide about 13 kWh per person per day. A different kind of solar panel uses the Sun's radiation to generate a voltage; these are called **photovoltaic** (PV) panels.

Hydroelectric power

Water heated by the Sun evaporates, and then falls as rain. Rain falling on high ground can be stored behind a dam and used to turn turbines in a **hydroelectric** power station as it flows downhill. The UK gets about 0.2 kWh per person per day from hydroelectric power. If more schemes were built this could rise to 1.5 kWh.

In the UK, placing PV panels on all south-facing roofs could provide 5 kWh per person per day

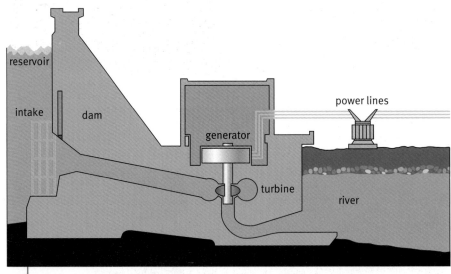

Water from the reservoir turns turbines, which turn the generator.

The Nant y Moch dam is part of a hydroelectric scheme in Wales. The power output from this scheme is 55 MW.

Wind power

Wind energy can be used to to turn a turbine, which drives an electricity generator. In the UK, a land-based collection of wind turbines (a **wind farm**) has an average output of 2 W per square metre. Wind farms covering the windiest 10% of the country could provide about 20 kWh per person per day. Wind farms built all around the coast of the UK could provide a further 48 kWh but there are major costs and engineering challenges when building at sea.

Power from waves and tides

The pull of gravity between the Earth and Moon affects the oceans. As the Moon orbits the Earth, and the Earth rotates, tides rise and fall. Ocean currents are driven by heating from the Sun and by Earth's rotation. Wind, currents, and tides combine to produce waves.

Water movement due to tides and waves can drive turbines. As the UK is surrounded by sea, **tidal power** could provide up to 11 kWh per person per day. There is a tidal energy convertor in Strangford Lough and the government has proposed building a tidal power station in the Severn Estuary.

Biofuels

Biofuels are renewable because they can be replaced quickly. Some biofuels could replace petroleum fuels for transport and all could be burned in thermal power stations. At best, biofuels could provide a total of 7 kWh per person per day in the UK.

Winds are driven by temperature differences in the atmosphere and by the Earth's rotation. The Whitelee wind farm near Glasgow will cover 55 km^2.

Miscanthus grass is grown for fuel. In the UK each square metre of crop yields about 0.2 W electricity.

In a geothermal power station energy from hot rocks is used to produce steam to drive turbines.

A Pelamis generator uses **wave power** to produce electricity. Pelamis machines along 500 km of Britain's Atlantic coast could produce 4 kWh per person per day.

Key words

- ✔ **renewable**
- ✔ **solar power**
- ✔ **thermal**
- ✔ **hydroelectric**
- ✔ **wind farm**
- ✔ **wave power**
- ✔ **tidal power**

Questions

1 A student says 'All our energy comes from the Sun.' Explain how this is true for the renewable energy sources mentioned on these pages.

2 List the drawbacks of each renewable resource.

3 What would be the maximum total energy in kWh per person per day that we could get from renewable sources in the UK? Suggest reasons why, in practice, the amount is likely to be much less.

Find out about

- why we need a National Grid
- why the National Grid uses very high voltages
- how transformers are used to alter the voltage of an electricity supply

Power stations are built close to their energy source or where there is plenty of cooling water. But that is not always where the electricity is needed. There are energy issues involved in the distribution of electricity, as well as in its generation.

National Grid

All the power stations in the UK are connected to the **National Grid**, which is used to distribute electricity to all the places where we want to use it. It does this by means of a network of long wires and transformers. The power sockets in your home are connected to the Grid. You don't just depend on the nearest power station.

When an electric current flows in a wire, it causes heating. Even if the heating is only slight, it means that energy is wasted rather than getting to the user. The National Grid covers large distances so losses due to heating can be significant.

High voltage

In the UK the domestic supply voltage is 230 V. This is high enough to give a fatal electric shock, but the voltage used over most of the National Grid is very much higher.

To distribute electric power at 230 V, very large currents would be needed. These currents would cause a lot of heating in the cables and so much energy to be wasted.

It is more efficient to use a high voltage to distribute electricity. The higher the voltage, the smaller the current needed for the same power output to the user. With a smaller current, less energy is lost due to heating in the wires.

Transformers

The voltage of an a.c. electricity supply can be altered using a **transformer**.

The high-voltage wires of the Grid are supported by tall pylons to keep them out of reach.

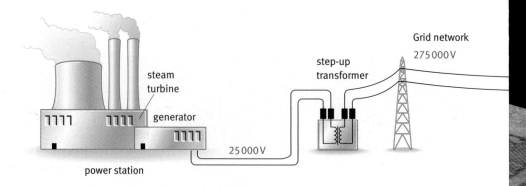

In a step-up transformer the output voltage is higher than the input voltage and the current is reduced. The National Grid uses step-up transformers to increase the voltage for transmission of electricity. A step-down transformer in the local substation reduces the voltage and the current is increased.

Key words

✓ **National Grid**
✓ **transformer**

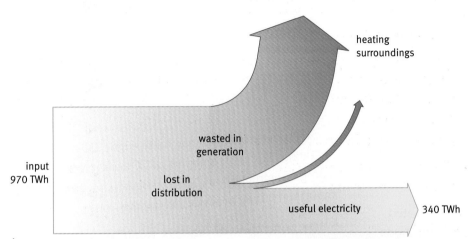

In 2008 the UK used about 970 TWh of energy to supply 340 TWh of electricity. 40 TWh was lost in the distribution process. 1TWh is one million MWh.

Questions

1 The UK National Grid is connected to the French grid. Suggest a reason for this.

2 Electricity power lines can be buried under the ground or carried by pylons. Suggest an advantage and a disadvantage of each method.

3 Use the Sankey diagram to calculate the overall efficiency of electricity generation and distribution in the UK.

4 Someone has written to their local paper complaining that large pylons and transformer substations are 'an ugly blot on the landscape' and asking why houses can't just be connected to the nearest power station with 'normal cables'. Write a letter explaining why the pylons and substations are needed.

This transformer is about to be installed in a substation.

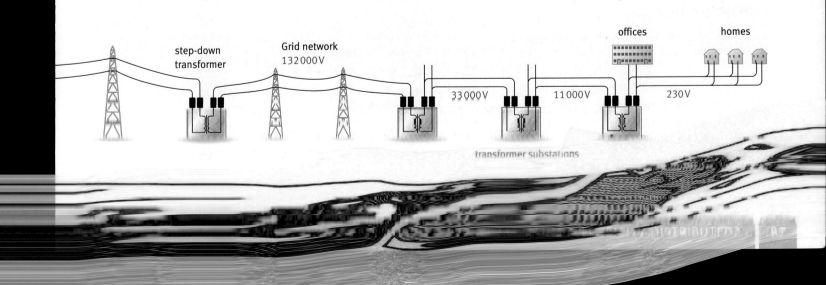

The energy debate

Nuclear

Find out about

- arguments for and against using various energy sources
- how to make your views known

YES

Nuclear power is the only energy source that can meet a substantial electricity demand. It releases no damaging carbon dioxide.

The best way to use world stockpiles of uranium and plutonium is as fuel in civilian reactors to generate electricity. Otherwise they remain available for making nuclear weapons.

UK nuclear power stations use tried-and-tested technology. Safety systems meet high standards. Waste disposal is a problem that can be solved.

NO

Nuclear power stations may release little CO_2 while operating. But large amounts of CO_2 are released during construction and decommissioning. Most importantly, they produce radioactive waste.

A new generation of reactors would take about a decade to build and cost roughly £2 billion each. No insurance company will cover their risks, during operation or decommissioning. The public will have to pay if anything goes wrong.

Renewable energy sources

NO

Renewable energy is unreliable. Winds don't always blow. The Sun doesn't always shine.

Renewable sources would not provide enough energy for this country.

No wind farm should be built where people live and work. Each wind turbine is a huge noisy machine.

Wave and tidal generators interfere with wildlife. Large hydroelectric schemes damage the countryside.

The main renewable energy sources are not in the same place as existing power stations. We would have to extend the National Grid with more power lines across the countryside.

YES

The UK should exploit its own energy sources and not rely on imports. Recent studies suggest that renewable energy sources could provide the UK with a reliable supply of electricity.

What we need is a full range of generators – very big to very small – at sites all around the country. A decentralised power system would be based on microgeneration. Installing wind generators and solar cells on the rooftops of many offices and homes will be relatively cheap and easy.

A life cycle assessment shows that power from the Sun and winds releases little CO_2.

Use less energy – for and against

YES

Energy consumption rises year by year. In your lifetime, you are likely to use as much energy as all four of your grandparents put together. Every energy saving you can make will help.

The government can help by:

- requiring new buildings to use less energy for heating and lighting
- providing grants to help householders install domestic combined heat and power systems
- ensuring that new appliances are energy efficient
- taxing fossil fuels more highly.

NO

It's all very well to dream of using less energy. But energy makes the world go round. It's essential to education, business, and pleasure. And everyone has a right to a good standard of living at home.

There isn't a simple answer to the energy question. There will have to be a mix of energy sources, as there is now. No single source can meet all our needs. And that leaves us with more questions:

- Who should make the decisions about which energy sources are in the mix – the energy companies, the government, or scientists?
- Who should have the last word when a few local people object to a new power station that will meet the needs of many more people?
- How do you weigh the benefits of 'clean' energy from wind turbines against the change in view across the hills?

Key word

✓ **decommission**

Questions

1. What does 'a sustainable supply of energy' mean?

2. The cost of decommissioning contributes to the price of electricity. It is much larger for nuclear power stations than for stations burning fossil fuels. Explain why.

3. Look at the arguments for and against each energy source.
   a. Draw balance diagrams for each option, listing statements on each side.
   b. Distinguish statements of fact from opinion, by putting a tick next to facts.
   c. Use the information in this book to add further statements to your diagrams.

4. Write a letter to your Member of Parliament expressing your views about future power stations. Use your answer to question 3 to make your letter persuasive and show that you have considered the issues.

The dismantling of Berkeley nuclear power station in Gloucestershire. Energy costs take in the whole life of a power station, from start to finish.

Science
Explanations

In this module you will learn about the uses of energy from fossil fuels and nuclear power and explore renewable sources of energy. You will also learn how electricity is generated and distributed.

You should know:

- that the demand for energy is continually increasing, and that this raises issues about the availability of sources and the environmental effect of using them
- the main primary energy sources
- why electricity is called a secondary energy source and why it is convenient to use
- which renewable energy sources are used for generating electricity
- that burning carbon fuels in power stations produces carbon dioxide
- that power is the rate at which energy is transferred
- that electrical energy transferred = power × time
- that electric power = voltage × current
- that joules and kilowatt-hours are both units of energy
- how to interpret and construct Sankey diagrams
- that efficiency of electrical appliances and power stations can be calculated using the equation:

$$\text{efficiency} = \frac{\text{energy usefully transferred}}{\text{total energy supplied}}$$

- that mains electricity is produced by generators, which contain coils of wire and spinning magnets
- that thermal power stations use a primary energy source to heat water to drive a turbine and generator, but that many renewable sources of energy drive a turbine directly
- how to label a block diagram showing the main parts of power stations
- that nuclear power stations produce radioactive waste, which emits ionising radiation
- the distinction between contamination and irradiation by a radioactive material
- how electricity is distributed through the National Grid at high voltage, although the main supply voltage to our homes is 230 V
- how to evaluate energy sources, using data where appropriate, in terms of:
  - where they are used (home, work place, or nationally)
  - factors that affect the choice of the source (the environment, economics, waste products produced)
  - the advantages and disadvantages of different non-renewable and renewable power stations (fossil fuel, nuclear, biomass, solar, wind, and water).

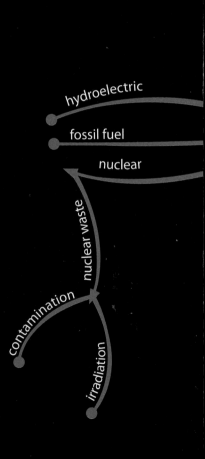

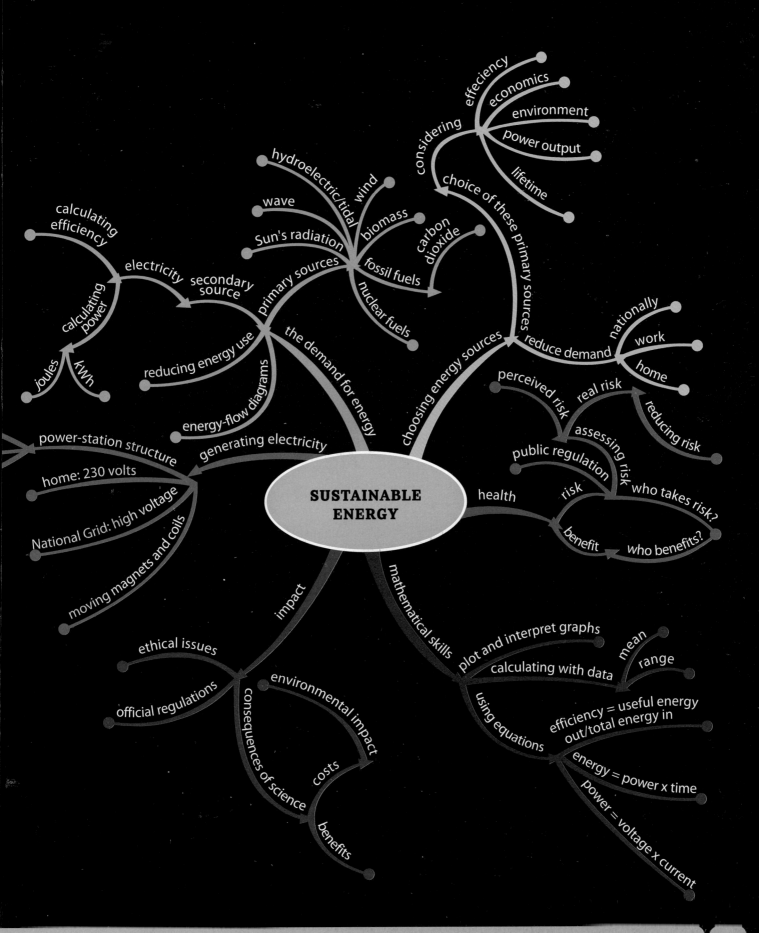

SUSTAINABLE ENERGY

considering
- effeciency
- economics
- environment
- power output
- lifetime

choice of these primary sources

primary sources
- hydroelectric/tidal
- wind
- wave
- Sun's radiation
- biomass
- fossil fuels → carbon dioxide
- nuclear fuels

secondary source → electricity
- calculating efficiency
- calculating power
  - joules
  - kWh

reducing energy use

the demand for energy

energy-flow diagrams

generating electricity
- power-station structure
- home: 230 volts
- National Grid: high voltage
- moving magnets and coils

choosing energy sources

reduce demand
- nationally
- work
- home

health
- risk
  - perceived risk
  - real risk
  - assessing risk
  - public regulation
  - reducing risk
  - who takes risk?
- benefit → who benefits?

impact
- ethical issues
- official regulations
- environmental impact
- consequences of science
  - costs
  - benefits

mathematical skills
- plot and interpret graphs
- calculating with data
  - mean
  - range
- using equations
  - efficiency = useful energy out/total energy in
  - energy = power x time
  - power = voltage x current

Ideas about Science

In addition to developing an understanding of the use and generation of electricity, it is important to assess the risks and benefits associated with the chosen methods of energy use, and to appreciate the issues involved in making decisions about the use of science and technology.

Everything we do carries some risk, and new technologies often introduce new risks. It is important to assess the chance of a particular outcome happening, and the consequences if it did, because people often perceive a risk as being different from the actual risk: sometimes less, and sometimes more. A particular situation that introduces risk will often also introduce benefits, which must be weighed up against that risk.

You should be able to:
- identify risks arising from scientific or technological advances
- suggest ways of reducing a given risk
- interpret and assess risk presented in different ways
- distinguish between real risk and perceived risk
- suggest reasons for given examples of differences between perceived and actual risk
- discuss how risk should be regulated by governments and other public bodies and explain why it may be controversial.

Science-based technology provides people with many things they value. However, some applications of science can have undesirable effects on the quality of life and on the environment. Benefits need to be weighed against costs.

In the context of the sustainable energy, you should be able to:
- identify the groups affected, and the main benefits and costs of a course of action for each group

- suggest reasons why different decisions on the same issue might be taken in different social and economic contexts
- identify examples of unintended impacts of human activity on the environment
- explain the idea of sustainability, and use it to compare the sustainability of different processes
- discuss the official regulation of the application of scientific knowledge
- in cases where an ethical issue is involved, say clearly what the issue is and summarise different views that may be held
- understand ethical arguments based on 'the greater good of the greater number' and that certain things are right or wrong whatever the circumstances.

Review Questions

1

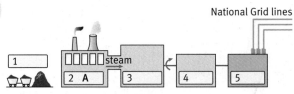

National Grid lines

a Copy and label the diagram of a coal-fired power station.

Put the letters **A**, **B**, **C**, **D**, and **E** in the correct boxes on the diagram. One has been done for you.

A	furnace
B	transformer
C	fuel
D	turbine
E	generator

b Power stations use a carbon-based fuel. Which greenhouse gas will definitely be produced when the fuel is burnt?

c Coal is a non-renewable energy source. Which two of the following are **renewable** energy sources that are used to generate electricity?

natural gas	**nuclear fuel**
wind power	**oil** **wave power**

2 The diagram shows the efficiency of a modern power station.

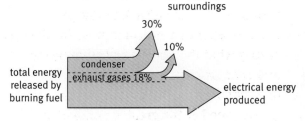

heating the surroundings

30%

10%

condenser

total energy released by burning fuel

exhaust gases 18%

electrical energy produced

a Use the diagram to calculate the efficiency of the power station in producing electrical energy.

b One way to waste less energy is to use the heat energy from the condenser to heat homes and businesses near the power station. Assuming **half** of the heat from the condenser can be used in this way, what is the efficiency of the power station in providing useful energy output?

3 An electric heater draws a current of 10 A from a 230 V power supply.

a Calculate the input power, in watts and in kilowatts, to the heater.

b Calculate the cost of using the heater for five hours, if one kilowatt-hour of electrical energy costs 8 p.

4 Many wind farms are being planned to generate electricity for Britain.
The pie chart below shows the various costs of setting up and operating a wind farms.

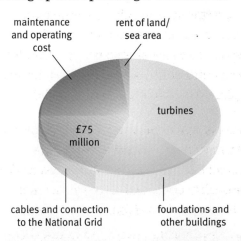

maintenance and operating cost

rent of land/ sea area

turbines

£75 million

cables and connection to the National Grid

foundations and other buildings

a Which one of the following is the best estimate of the cost of the turbines?

£40 million	**£75 million**
£150 million	**£200 million**

b Which one of the factors costs £90 million?

P4 Explaining motion

Why study motion?

Humans have always been interested in how things move and why they move the way they do. Motion is such an obvious part of our everyday lives that we cannot really claim to know very much about the natural world if we cannot explain and predict how objects move.

What you already know

- Speed is calculated by the distance covered divided by time.

- An unbalanced force changes the motion of an object.

- The weight of an object is due to the gravitational force between the Earth and the object.

- The air resistance on a moving object depends on its shape and its speed.

Find out about

- forces always arising from an interaction between two objects

- friction and reaction of surfaces

- instantaneous and average speed, velocity, and acceleration

- the idea of momentum, and how the momentum of an object changes when a force acts on it

- everyday examples of motion, including the principles on which traffic-safety measures are based

- gravitational potential energy and kinetic energy.

The Science

One tantalising thought has always driven people who have studied motion – is it possible that every example of motion we observe can be explained by a few simple rules (or laws) that apply to everything? Remarkably the answer is 'yes'. And these laws are so exact and precise that they can be used to predict the motion of an object very accurately.

Ideas about Science

A scientific explanation cannot be deduced by just looking at the data – it needs someone to think creatively to explain the observations. Many people had tried to describe and explain how things move. To write his laws of motion Isaac Newton built on the ideas of those who had come before. But making the link between an apple falling on the ground and the Moon orbiting the earth required a leap of imagination.

Find out about

- ✔ **how forces arise when two objects interact**
- ✔ **contact and action-at-a-distance forces**

What makes things move the way they do? And what makes things stop, or stay still? To start something moving, we have to push or pull it. To stop a moving object, we have to exert a **force** on it, against its direction of motion.

Where do forces come from?

Look at the photograph of a firework rocket exploding. The burning sparks move rapidly out from the centre, where the chemicals in the rocket have exploded. Notice that the starburst is symmetrical. For every moving spark, another spark moves in the exact opposite direction. This tells us something very important about forces. They always come in pairs.

Let's consider a simpler example, where there are two moving objects. Sophie and Sam stand in the centre of an ice rink. What happens if Sophie gives Sam a gentle push?

The chemical reaction inside the firework produces forces that send the burning fragments out equally in all directions, producing a sphere of sparks.

An exploding firework rocket. For every moving spark, there is another spark moving in exactly the opposite direction.

The answer is that both of them will move. When Sophie exerts a force on Sam by pushing him, she experiences a force herself. Sam exerts a force on Sophie – not by pushing but just by being there as an obstacle to push against. The same is true if they stand a distance apart holding a rope. If either one pulls, both will start to move together.

This tells us something very important about forces:

- Forces always arise from an **interaction** between two objects. So forces always come in pairs. The two forces in an **interaction pair** are:
  - equal in size
  - opposite in direction.

This is always true. And it does not depend on the size or strength of the two people involved. Another important thing to notice is that

- the two forces act on different objects.

In this example, one force of the pair acts on Sam and the other on Sophie.

force exerted by Sam on Sophie   force exerted by Sophie on Sam

Forces always arise in pairs. Here Sophie pushes Sam and also experiences a force herself in return.

Two kinds of interaction

Forces arise from interactions between pairs of objects. When they are caused by two objects touching, we call them **contact forces**. Contact forces exist only while the objects are actually touching. As soon as the objects separate, the forces stop. The two objects may, of course, keep moving. We will come back to this later in this chapter.

There is a second kind of interaction between objects, called **action-at-a-distance**. One common example is magnetism. The two ring magnets in the diagram are repelling each other. Both threads are at an angle – because both magnets experience a force. The same is true when magnets attract. If you hold a fridge magnet close to the door, the magnet experiences a force towards the door – and the fridge experiences a force towards the magnet.

Gravity is another example of action-at-a-distance. An apple falls from a tree because of the force exerted on it by the Earth, pulling it downwards. But gravity is an attraction between two objects. The apple also exerts an equal and opposite force on the Earth! This does not have any visible effect, however, because the Earth is so massive.

Unlike contact forces, action-at-a distance forces act all the time, even when the two interacting objects are apart. They get weaker as the distance between the objects increases.

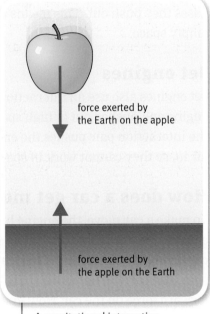

force exerted by the Earth on the apple

force exerted by the apple on the Earth

A gravitational interaction – again forces always arise in pairs.

Both the fridge magnet and the door experience a force. Action-at-a-distance forces always come in pairs.

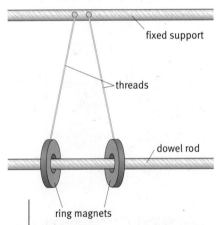

fixed support

threads

dowel rod

ring magnets

These two ring magnets are repelling each other. Notice that both magnets are being pushed aside.

Questions

1 List four examples of interaction pairs of forces mentioned on these pages.

2 What three things are always true about interaction pairs?

3 How could you modify the apparatus in the top diagram to show that *attraction* forces between magnets also arise in pairs? Sketch how you would set it up, and write down what you would expect to see.

Key words

- ✓ **force**
- ✓ **interaction**
- ✓ **interaction pair**
- ✓ **contact forces**
- ✓ **action at a distance**

Find out about

- **the forces that enable people and vehicles to get moving**
- **rockets and jet engines**

To get something moving, an interaction pair of forces is always involved. Here are some examples.

Rockets

As its fuel burns, a rocket pushes out hot gases from its base. The rocket exerts a large force downwards on these gases. The other half of this interaction pair is the force exerted on the rocket by the escaping gases. This pushes the rocket upwards.

The photograph on the left shows one of the most famous rocket launches: the *Apollo 11* mission to land the first humans on the Moon. The interaction pair of forces is shown on the photo.

Rockets carry with them everything they need to make the burning gases they push out. This means that they can work anywhere, including empty space.

Jet engines

Jet engines also use an interaction pair of forces. Air is drawn into the engine and pushed out at high speed from the back. The other force of the interaction pair pushes the engine forward. Jet engines need to draw air in, so they cannot work in space.

How does a car get moving?

To make a car move, the engine has to make the wheels turn. This causes a forward force on the car. To understand how, think first about a car trying to start on ice. If the ice is very slippery, the wheels will just spin. The car will not move at all. The spinning wheels produce no forward force on the car. Now imagine a car on a muddy track. The rally car below is throwing up a shower of mud as it tries to get going.

force exerted on the rocket

force exerted on the hot gases

The start of the longest journey humans have made so far – to the Moon. A huge force is needed to push a rocket like this upwards. It is provided by the hot exhaust gases, which are formed by burning the fuel.

As it rotates, the wheels exert a force backwards on the ground – with dramatic results in this case!

You can see that there is an interaction between the wheels and the ground. The wheels are causing a backwards force on the ground surface. This makes the mud fly backwards. Mud, however, moves when the force is quite small. The other force of the interaction pair is the forward force on the car. It is equal in size. So it is also small – and not big enough to get the car moving.

Now imagine a good surface and good tyres, which do not slip. Again, the engine makes the wheel turn. It pushes back on the road. Friction between the wheel and the road surface stops it slipping so it exerts a very large force backwards on the road surface. So the other force of the interaction pair is the same size. This large forward force gets the car moving.

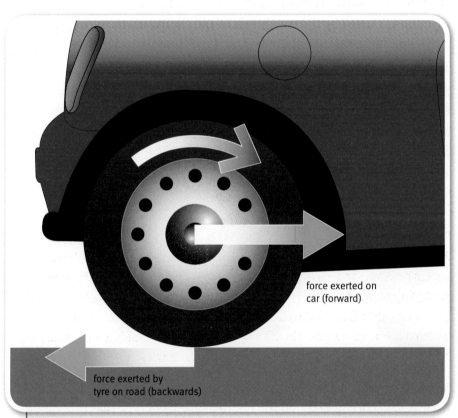

force exerted on car (forward)

force exerted by tyre on road (backwards)

If the tyre grips the road and does not spin, the second force of the interaction pair results in a large forward force on the axle. This pushes the car forward.

Walking

When you walk, you push back on the ground with each foot in turn. The ground then pushes you forward. You are not usually aware of this. When you walk across a floor, it does not feel as though you are pushing backwards on it. You only become aware of the importance of this interaction when the surface is slippery – for example, when you try to walk on an icy surface, where there is not enough friction. Because you cannot push it back, it is unable to move you forward.

Questions

1 Jet engines are suitable for aircraft but not for travel in space. Explain why. How do rockets overcome this problem?

2 A boat propeller pushes water backwards when it spins round. Use the ideas on these pages to write a short paragraph explaining how this makes the boat move forward. Draw a diagram and label the main forces involved, to illustrate your explanation.

3 Sketch a matchstick figure walking. Mark and label the interaction pair of forces on the foot in contact with the ground.

Icy surfaces are difficult to walk on. Your foot cannot get a grip to push back on the surface, so the surface does not push you forwards.

Friction is the interaction between two surfaces when they slide over each other – or when they are pushed to try to make them slide over each other. There is a friction force on *both* objects involved. As always, the two forces of the interaction pair are equal in size and opposite in direction. The friction force on each object acts in the direction that would help to prevent it sliding over the other.

Friction enables cars and people to get moving. If friction didn't exist, walking would be impossible and wheeled vehicles could not move. But what causes friction and how does it work?

What causes friction?

Friction is caused by the roughness of the sliding surfaces. Even surfaces that seem smooth have quite large humps and hollows if you look at them under a microscope.

So, at a microscopic level, two 'smooth' surfaces in contact are not really touching everywhere. And some of the bumps on one surface will fit into hollows on the other. As one slides across the other, these bumps collide. This causes a force that resists the sliding.

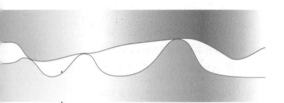

The bumps and hollows on each surface cause a force, as one object slides (or tries to slide) across the other.

×1080

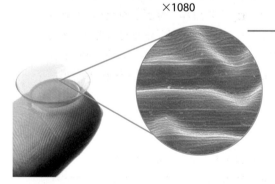

Even surfaces that appear smooth are really quite rough. At the microscopic level they have humps and hollows. The photograph on the right shows the surface of a contact lens magnified 1080 times.

Friction between an object and the floor results in a pair of opposite forces. The arrows show the direction of the forces exerted on the foot and the floor. In reality, the bumps and hollows in the two surfaces are very much smaller.

When you walk, it's as if your shoes have corrugated soles – and the floor is also corrugated – on the microscopic level. The diagram on the left exaggerates this in order to show the effect more clearly. Pushing back with your foot will cause a backwards force on the ground, and a forwards force on your foot.

Questions

1 List three everyday situations in which we try to reduce friction, and three where we try to make friction as large as possible.

2 Use the ideas on these pages to write short explanations of the following observations:

a We can reduce the friction between two surfaces by putting oil on them.

b It is easier to push a box across the floor when it is empty than when it is full.

Friction – a responsive force

What size is the friction force between two surfaces, and what does it depend on? Think about the forces involved as Jeff tries to push a large box along a level floor.

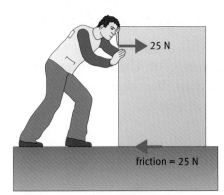

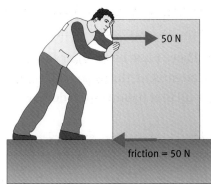

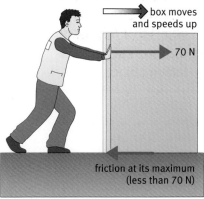

box moves and speeds up

1. Jeff pushes the box with a force of 25 N, to try to slide it along. It does not move. The friction force exerted by the floor on the box is 25 N. This exactly balances Jeff's push.

2. Jeff then pushes harder, with a force of 50 N. The box still does not move. The friction force exerted by the floor on the box is now 50 N. Again, this balances Jeff's push.

3. Jeff pushes harder still, exerting a force of 70 N. The box starts to move. 70 N is bigger than the maximum friction force for this box and floor surface.

The size of the friction force depends on the size of the external force applied by Jeff, up to a certain limit. If Jeff's push is below this limit, friction matches it exactly. The two forces cancel each other out and the box doesn't move. But if Jeff's force is above this limit, the box will move. So what determines this limit? It depends on the weight of the box, and on the roughness of the two surfaces in contact.

Adding forces

The discussion above used an idea that may seem obvious:

- If there is a force acting on an object, but it is not moving, then there must be another force balancing (or cancelling out) the first one.

If the forces acting on an object balance each other, we say they add to zero. Adding several forces that act on the same object is straightforward. But you must take the direction of each force into account. The sum of all the forces acting on an object is called the **resultant force**. The diagrams on the right show some examples.

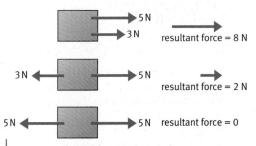

To find the resultant force acting on an object, you add the separate forces. You must take account of their directions.

Key words
- ✓ **friction**
- ✓ **resultant force**

Find out about

- ✓ **how a surface exerts a reaction force on any object that presses on it**
- ✓ **the forces acting on a falling object**

If you hold a tennis ball at arm's length and let it go, it immediately starts to move downwards. There is a force acting on the tennis ball. This force is the pull exerted on it by the Earth. It is due to the interaction known as gravity.

But if you put a tennis ball on a table so that it does not roll about, it does not fall. The force of gravity has not suddenly stopped or been switched off. There must be another force that cancels it out. The only thing that can be causing this is the table. The table must exert an upward force on the ball that balances the downward force of gravity.

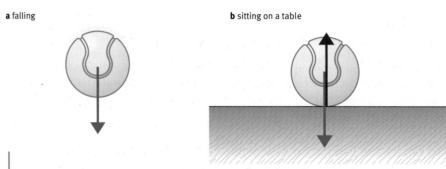

a falling   **b** sitting on a table

The forces acting on a tennis ball **a** falling and **b** sitting on a table.

How can a table exert a force?

Although it may seem strange, tables can and do exert forces. To understand how, imagine an object, like a school bag, sitting on the foam cushion of a sofa. The bag presses down on the foam, squashing it a bit. Because foam is springy, it then pushes upwards on the bag, just like a spring. Like a spring, the more it is squeezed, the harder it pushes back. So the bag sinks into the foam until the push of the foam on it exactly balances the downward pull of gravity on it.

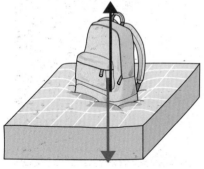

force exerted by the cushion on the bag (reaction)

force exerted by the Earth on the bag (due to gravity)

The bag squeezes the foam until the upward force of the springy foam on the bag exactly balances the downward gravity force on the bag.

The same thing happens, though on a much smaller scale, when the bag sits on a table top. A table top is not so easily squeezed as a foam cushion. But it *can* be squashed. This is not visible to the naked eye, however. We call this upward force that a surface exerts when something presses on it the **reaction** of the surface.

The size of the reaction force depends on the downward force that is causing it. The surface distorts just enough to make the reaction force balance the downward force on the object due to gravity. The resultant force on it is zero.

There is, of course, a limit to this. If the downward force exerted on a table is bigger than it can take, the table top will break! But up to this limit, the reaction matches the downward force.

Walls can push too!

Any surface can exert a reaction force, not just horizontal ones. The diagram on the right shows Deborah, a roller-skater. When she pushes on the wall, it pushes back on her. She immediately starts to move backwards. Deborah's push squashes the wall where her hands touch it. Although we cannot see any distortion, this part of the wall is compressed. Like a spring, it exerts an equal force back on Deborah's hands.

Freefall

So how does a tennis ball actually move, when it is falling? If we look closely, we see that it picks up speed steadily from the moment it is released. The multi-flash photograph on the right shows this. The time interval between flashes is constant. As the ball falls, it moves further between one flash and the next. Its speed increases. It accelerates.

If we drop an object that is light relative to its size, like a cupcake case, we notice that something different happens. It speeds up at first, then falls at a steady speed. Why the difference? The reason is that **air resistance** has a bigger effect on it. Air resistance is a force that arises when anything moves through the air. The faster an object moves, the bigger the air resistance force on it becomes. A falling light object quickly reaches the speed at which the air resistance force on it balances the force of gravity. It then continues to fall at a steady speed.

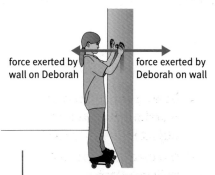

force exerted by wall on Deborah

force exerted by Deborah on wall

Deborah pushes against the wall. The other force, in the interaction pair, starts her moving away from the wall.

Freefall – a falling ball gets steadily faster as it falls.

Questions

1 What happens to a 'hard' surface when something sits on it? Is it really as hard as it seems?

2 Imagine the bag in the diagram opposite hanging from a string. The string must be exerting an upward force on the bag, equal to the downward force of gravity on it. How does the string exert this upwards force? Use the ideas on these pages to suggest an explanation.

3 Draw a diagram to show the forces acting on a cupcake case when it is falling.

Key words
- ✓ reaction (of a surface)
- ✓ air resistance

Find out about

- ✔ **how to calculate the speed of a moving object**
- ✔ **how to calculate the acceleration of a moving object**

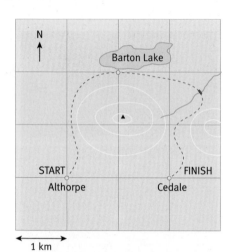

The distance from start to finish is 6 kilometres. But the displacement at the finish is 2 kilometres east of the start.

In the previous sections you have seen how forces arise in interactions, and that there are some links between forces and motion. To explore these more fully, we need to be able to describe the motion of an object more clearly.

Distance and displacement

One obvious question to ask about the motion of an object is – how far did it move? But this could mean two different things. A group of walkers follow the route shown in the map. When they finish their walk, how far have they gone? The **distance** along the trail is 6 kilometres. But another answer to the question is that they have gone 2 kilometres east. This is called their **displacement**. The displacement is the straight-line distance and direction from the starting point. For walkers, distance is usually the more important quantity to know, as it helps you to judge how long the walk will take. But for sailors displacement is often more useful and important.

Speed

To find the speed of an object, you measure the time it takes to travel a known distance. You can then calculate its **average speed**, using the equation:

$$\text{average speed (metres per second, m/s)} = \frac{\text{distance travelled (metres, m)}}{\text{time taken (seconds, s)}}$$

The **instantaneous speed** of an object is the speed at which it is travelling at a particular instant. To estimate the instantaneous speed of an object, we measure its average speed over a very short distance (and hence over a very short time interval). The shorter we make this time interval, the less likely it is that the speed has changed much during it. On the other hand, if we make it very short, it is harder to measure the distance and the time accurately.

Question

1 The walkers use a straight farm track from Cedale back to Althorpe to pick up their cars. When they get back to their cars:
   a what is the total distance they have walked?
   b what is their total displacement?

A car's speedometer measures its average speed over a short time interval, giving a good indication of the instantaneous speed.

Worked example

An athlete runs a 100 metre race. The diagram shows her position at 1 second intervals during the race. She runs 100 metres in 12.5 seconds.

$$\text{average speed} = \frac{100 \text{ m}}{12.5 \text{ s}} = 8 \text{ m/s}$$

But she didn't run at 8 m/s for the whole race,

sometimes she ran more quickly, sometimes she ran more slowly; this is why we call this value the *average* speed.

As she crossed the finish line, she ran approximately 10 metres in one second, so her *instantaneous* speed as she crossed the finish line was about 10 m/s.

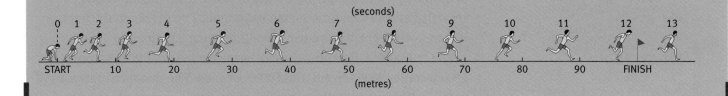

Velocity and acceleration

People often use the words 'speed' and 'velocity' to mean the same thing. The **velocity** of an object, however, also tells you the direction in which it is moving. So, for example, a cyclist is pedalling at 8 m/s along a road that runs due west. Her instantaneous speed is 8 m/s, but her instantaneous velocity is 8 m/s in a westerly direction.

In everyday language, if the speed of an object is increasing, we say that it is accelerating. Drivers are often interested in the **acceleration** of their car. This might be stated as 0–60 miles per hour in 8 seconds. In situations like this, where the direction of motion does not matter, we can use the equation:

$$\text{acceleration} = \frac{\text{change of speed}}{\text{time for the change to occur}}$$

A more complete definition of acceleration that applies to all situations is:

$$\text{acceleration (metres per second) per second, (m/s}^2) = \frac{\text{change of velocity (metres per second, m/s)}}{\text{time taken for the change (seconds, s)}}$$

Questions

2 Calculate the average speed of the athlete over the first 20 metres of the race. Explain how this shows that she accelerated during the race.

3 During which time interval do you think the athlete was running fastest? Explain your answer.

4 A high-performance car can accelerate from 0–27 m/s (0–60 mph) in 6 seconds. Calculate the average acceleration of the car in m/s².

Key words

- ✔ distance
- ✔ displacement
- ✔ average speed
- ✔ instantaneous speed
- ✔ velocity
- ✔ acceleration

Find out about

✔ **how graphs can be used to summarise and analyse the motion of an object**

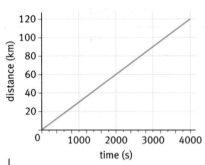

Distance–time graph for a car journey along the motorway.

Graphs are very useful for summarising information about the motion of an object over a period of time. It would take many words to describe the motion of an object as clearly and exactly as a graph can do. A graph provides information about the moving object at any instant. From its shape and slope, we can quickly see how the object moved.

Distance–time graphs

A **distance–time graph** shows the distance a moving object has travelled at every instant during its motion. The graph on the left is a distance–time graph for a car travelling along a motorway. Use the graph to find the distances the car has travelled after 1000 seconds, 2000 seconds, 3000 seconds, and 4000 seconds. The distance increases steadily. The speed of the car is constant. The constant **slope** of the distance–time graph indicates a steady speed.

The second graph below shows a more complicated journey. This is a cycle ride taken by Vijay during his school holidays. The graph has four sections. In each section the slope of the graph is constant. This means that Vijay's speed is constant during that section of the ride.

- In the first hour, Vijay travels 15 kilometres. His speed is a steady 15 km/h.
- In the second hour he travels only 5 kilometres, because the route goes steeply uphill. The shallower slope indicates a lower steady speed.
- In the third section (from 2.0 to 2.5 hours), his distance travelled does not change. He has stopped. This is what a horizontal section of a distance–time graph means.

Questions

1 How far does Vijay travel during the final section of his cycle ride (from 2.5 to 3.5 hours)? So what is his speed during this section?

2 Vijay was travelling faster in this final section of the ride than in the second (uphill) section, but not as fast as in the first section. Explain how you could tell this, just by looking at the graph.

I travel 5 km between these two points

So my speed is $\frac{distance}{time} = \frac{5\ km}{1\ hour}$ = 5 kmph

It takes me 1 hour

Vijay's cycle ride.

This distance–time graph for Vijay's journey is not very realistic. In a real journey, the speed would change gradually rather than suddenly.

Look at the distance–time graph on the right. It shows a car journey. The gradient tells you the speed of the car. If the gradient gets steeper, the car is speeding up, or accelerating. If the gradient is getting less steep, then the car's speed is decreasing.

Displacement–time graphs

If an object moves in a straight line, we can also draw a **displacement–time graph** of its motion. The graph shows how the size of the object's displacement changes with time. If the direction of motion does not change, the displacement–time graph is exactly the same as the distance–time graph. But if the direction of motion changes, the two graphs are different. The graphs below show the motion of a ball thrown vertically upwards from the moment it leaves the thrower's hand until the moment it arrives back again.

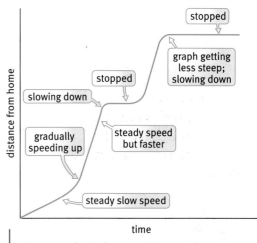
A more realistic distance–time graph.

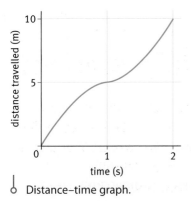
Distance–time graph.

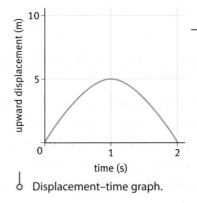
Displacement–time graph.

Two graphs showing the motion of a ball thrown up in the air.

Questions

3 Look at the graphs showing the motion of the ball thrown in the air.
   a Describe how the slope of both graphs changes over the first second of the ball's motion. What does this mean is happening to its speed? Is this what you would expect?
   b What is the speed of the ball after 1 s of its motion? Explain how you can tell this from the graphs.
   c How high does the ball go?
   d Explain why the displacement–time graph turns down after 1 s, whereas the distance–time graph continues to go up.

4 Roberta, an athlete, trains by jogging 20 m at a steady speed, then sprinting 20 m at a faster speed. She repeats this five times. Sketch a speed–time graph of her motion during a training session.

5 Think about the motion of a ball thrown upwards. Its speed gets steadily less on the way up and increases steadily again on the way down. Draw its:
   a speed–time graph
   b velocity–time graph

from the moment it leaves your hand until the moment it lands back in your hand.

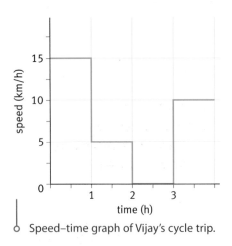

Speed–time graph of Vijay's cycle trip.

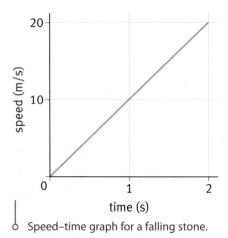

Speed–time graph for a falling stone.

Key words

✔ **distance–time graph**
✔ **displacement–time graph**
✔ **slope**
✔ **velocity–time graph**

Speed–time graphs

A speed–time graph shows the speed of a moving object at every instant during its journey. The speed–time graph for Vijay's trip would look like the graph on the left. A steady speed is now shown by a straight horizontal line.

Again, the sudden changes of speed shown on this graph are not realistic. It would take time for the speed of a moving object to change. A more realistic speed–time graph would have smoother, more gradual changes from one speed to another.

The second speed–time graph on the left is for a stone being dropped from a high bridge into a river. It has no horizontal sections, so the speed of the stone is changing all the time. The constant slope of the speed–time graph shows that the speed is changing at a steady rate. It has steady (or uniform) acceleration downwards.

From this speed–time graph, we can calculate the acceleration of the stone. Its change of speed is 20 m/s (from 0 to 20 m/s) in 2 seconds. So its acceleration is:

$$\frac{\text{change of speed}}{\text{time taken for the change}} = \frac{20 \text{ m/s}}{2 \text{ s}} = 10 \text{ m/s}^2$$

Velocity–time graphs

Velocity (as explained earlier) has a direction as well as a size. It is not possible to show both size and direction on a graph. However, for an object that is moving in a straight line, we can draw a graph of *the size of its velocity* against time. This is called a **velocity–time graph**.

Look at the graph below. It shows the motion of Karl's skateboard up and then down a slope. At first Karl is travelling at 10 m/s. He gradually slows down until his velocity is zero. But then the line of the graph keeps on going down. The velocity becomes negative. This may seem strange, but it is used to show that the skateboard's direction of motion has changed. Karl is now travelling in the opposite direction.

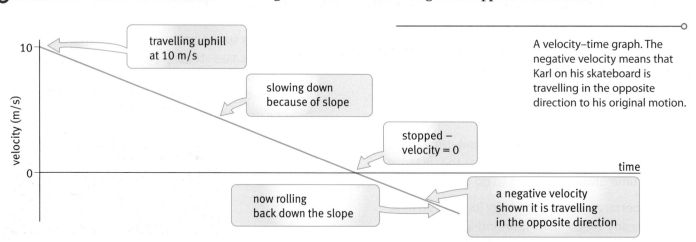

A velocity–time graph. The negative velocity means that Karl on his skateboard is travelling in the opposite direction to his original motion.

travelling uphill at 10 m/s

slowing down because of slope

stopped – velocity = 0

now rolling back down the slope

a negative velocity shown it is travelling in the opposite direction

You can now *describe* motion clearly; let's move on to *explain* motion. The key idea for explaining motion is force. If you know all the forces acting on an object, you can explain its motion.

Forces change motion

A force changes the motion of an object. Imagine a small toy cart, with smooth, well-oiled wheels, sitting on a level table top. Someone (or something) exerts a constant force on it towards the right. The cart will immediately start to move – and its speed will keep on increasing as long as the force continues.

To study the effect of a constant force in more detail, you can use the arrangement shown in the second diagram. The force of gravity on the hanging weight is constant, so the string exerts a steady force on the cart. You could then measure the speed of the cart at different time intervals after this pulling force is applied. The speed–time graph is a straight line. The speed of the cart increases steadily with time. A constant force gives the cart a constant acceleration.

Freefall is another example of motion with constant acceleration. In the flash photograph of a falling ball in Section D the gaps between the positions of the ball keep increasing as it falls. It gets steadily faster. The speed–time graph of a falling stone in Section F summarises the motion of a freely falling stone. The motion of a falling object is due to the gravitational force exerted on it by the Earth. This is a constant force on the object, so it causes its speed to increase steadily, a constant acceleration.

Momentum

A force acting on an object causes a *change* in its motion. To explore the connection between force and change of motion, let's consider a situation where two objects experience forces of exactly the same size. This happens in an interaction, for example, when two stationary objects spring apart.

In the diagrams on the next page, a spring-loaded cart is released, making the two objects move apart.

In the diagrams on the next page

Find out about

✔ **momentum**
✔ **the link between change of momentum, force, and time**

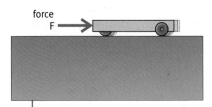

The speed of the cart keeps increasing as long as the force acts on it.

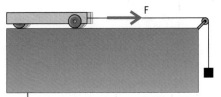

The string exerts a steady force on the cart.

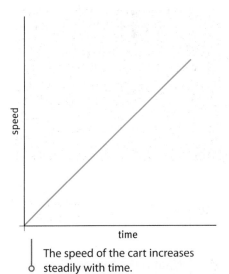

The speed of the cart increases steadily with time.

Key words

✔ **momentum**
✔ **change of momentum**

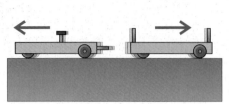

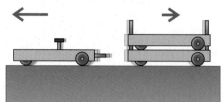

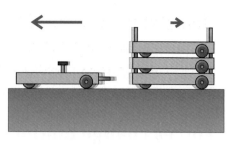

1. The left hand cart is spring loaded. If you put a second cart in front of it and release the spring, both move, in opposite directions. The interaction causes two forces, one on each cart. If the cart are identical, they both move with the same speed.

2. Here one cart is twice as heavy as the other. When the spring is released, both move. But the heavier one has only half the speed of the lighter one.

3. If we repeat this with a stack of three carts, we again find that both objects move. The speed of the stack of three carts is now one-third of the speed of the single cart.

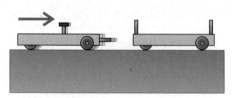

The momentum of the moving cart before the collision is the same as the momentum of the two carts after the collision. In all collisions, the total momentum is the same before and after – another indication that momentum is an important quantity for understanding motion.

In all of the interactions above, the size of the quantity (mass × velocity) is the same for both of the objects involved. This suggests that (mass × velocity) is a useful and important quantity. It is called the **momentum** of the moving object.

$$\text{momentum} = \text{mass} \times \text{velocity}$$
$$(\text{kg m/s}) \quad (\text{kg}) \quad (\text{m/s})$$

The faster an object is moving, the more momentum it has. A heavy object has more momentum than a lighter object moving at the same speed.

Force and change of momentum

In an interaction (like the three in the diagram above), the **change of momentum** is the same size for both objects. Two other things are also the same for both objects:

- the resultant force acting on the object (because the two forces in an interaction are always equal in size)
- the time for which it acts (the duration of the interaction, which has to be the same for both).

This suggests that there is a link between the force acting on an object, the time for which it acts, and the change of momentum it produces:

$$\text{change of momentum} = \text{resultant force} \times \text{time for which it acts}$$
$$(\text{kg m/s}) \quad\quad (\text{N}) \quad\quad (\text{s})$$

Questions

1. What is the momentum of:
   a. a skier of mass 50 kg moving at 5 m/s?
   b. a netball of mass 0.5 kg moving at 3 m/s?
   c. a whale of mass 5000 kg swimming at 2 m/s?

2. When a force makes an object move, which two factors determine the change of momentum of the object?

3. Calculate the change of momentum for:
   a. a force of 40 N acting for 3 s
   b. a force of 200 N acting for 0.5 s
   c. a force of 3 N acting for 50 s.

4. Use ideas about force and momentum to explain how the burning gases make a rocket or jet engine move.

Using the change of momentum equation

We can use the change of momentum equation to calculate the force applied.

Kicking a football

When a footballer takes a free kick, there is an interaction between his foot and the ball. His foot exerts a force on the ball.

This force lasts for only a very short time, the time for which the foot and the ball are actually in contact. After that, the player's foot can no longer affect the motion of the ball. The kick has not given the ball some 'force' but it has given it some momentum.

Taking a free kick. The interaction between the footballer's foot and the ball causes a change of momentum.

Worked example

How big is the force?
A football has a mass of around 1 kg. A free kick gives it a speed of 20 m/s. What is its momentum?

The football's momentum is given by the equation

$$\text{momentum} = \text{mass} \times \text{velocity}$$
$$= 1 \text{ kg} \times 20 \text{ m/s}$$
$$= 20 \text{ kg m/s}$$

It started with speed zero so the ball's change of momentum during the kick is 20 kg m/s.

The contact time when a football is kicked is around 0.05 s.

Estimate the force on the ball during the kick.

Use the equation

$$\text{change of momentum} = \text{force} \times \text{time it acts}$$
$$20 \text{ kg m/s} = \text{force} \times 0.05 \text{ s}$$

Dividing both sides by 0.05 s, you get

$$\frac{20 \text{ kg m/s}}{0.05 \text{ s}} = \text{force} = 400 \text{ N}$$

So the average force on the ball during the kick is 400 N. The force will not be constant during the time the ball is in contact. It will rise to a maximum (more than 400 N) and fall again to zero in 0.05 s.

Newton's Laws

This link between force, time, and change of momentum was proposed by Isaac Newton in 1687 in his famous book, the *Principia* (the *Mathematical Principles of Natural Science*). Newton did not use the word 'momentum' but wrote instead about an object's 'amount of motion' and how a force acting for a time could change it. The way he uses the word makes it clear that he is talking about the quantity we now call momentum.

Newton's hypothesis has been supported by many observations and measurements since then. There are all kinds of examples of motion, from everyday objects to the motion of the planets and stars. We now accept it as a reliable rule for explaining and predicting the motion of anything, except the very small (atomic scale) and the very fast (approaching the speed of light).

Find out about

✓ **safety features in cars**

This driver, Hybrid 111, has experienced many crashes. It is packed with sensing equipment to record forces on different areas, like the head, chest, and neck. Each dummy costs more than £100,000 to build.

Key word

✓ **risk**

Questions

1 When you jump down from a wall or ledge, it is almost automatic to bend your knees as you land. Use the ideas on this page to explain why this reduces the risk of injury.

2 In railway stations, there are buffers at the end of the track. These are a safety measure, designed to stop the train if the brakes fail. Use the ideas on this page to explain how buffers would reduce the forces acting on the train and on the passengers.

Cars today are much safer to travel in than cars 10 or 20 years ago. As a result of crash tests like the one shown on the left, designs have changed and are still changing.

If a car is travelling at 100 km/h, the driver and passengers are also travelling at that speed. If the car comes to a very sudden stop, due to a collision, the occupants will experience a very sudden change in their momentum. This could cause serious injury.

Crumple zones

Look at the diagram below. Which car would be safer in a collision? The answer may not be so obvious. You need to think about the change of momentum during the collision and the time the collision lasts.

a

Would you be safer in car **a** or car **b**?

b

The momentum of a moving car depends on:
• its mass
• its velocity.

In a collision, the car is suddenly brought to a stop. Its momentum is then zero. The size of the average force exerted on the car during the collision depends on the time the collision lasts:

change of momentum = average force × time for which it acts

The bigger the time, the smaller the average force – for the same change of momentum. This is why cars are fitted with front and rear crumple zones, with a rigid box in the middle. They are designed to crumple gradually in a collision. This makes the duration of the collision (the time it lasts) longer. This then makes the average force exerted on the car less.

The passengers inside the car also experience a sudden change of momentum. They were moving at the same speed as the car and are suddenly brought to a stop. A force exerted on their bodies (by whatever they come into contact with) causes this change of momentum. The longer time it takes to change a passenger's speed to zero, the smaller the average force experienced.

Seat belts and air bags

Some people think that seat belts work by stopping you moving in a crash. In fact, to work, a seat belt actually has to stretch. Seat belts work on the same principle as crumple zones. They make the change of momentum take longer. So the average force that causes the change is less. A crash helmet works in the same way. As the helmet deforms, it increases the time it takes for your head to stop, so the force is less.

With a seat belt, the top half of your body will still move forward, and you may hit yourself against parts of the car. Air bags can help to cushion the impact. Again, they reduce your momentum over a longer time interval so that the average force you experience is less.

Could you save yourself?

Some people think they could survive a car accident without a seat belt, especially if they are travelling in the back seats. A car is travelling at 50 km/h (or roughly 14 m/s). Without a seat belt, the driver would hit the steering wheel and windscreen about 0.07 seconds after the impact. Back-seat passengers would hit the back of the front seats at roughly the same time. As your reaction time is typically about 0.14 seconds, this would all happen before you even had time to react. Even if you could react in time, the force needed to change your speed from 14 m/s to zero in 0.1 s is larger than your arms or legs could possibly exert.

Safety and risk

Car safety features, such as seat belts, are designed to reduce the **risk** of injury in a car accident. Travel can never be made completely safe. We cannot reduce the risk to zero. But many studies, in different countries, have shown that wearing seat belts greatly reduces the risk of serious injury. Many countries have regulations requiring drivers and passengers to use seat belts.

1 0.00 s

2 0.05 s

3 0.10 s

4 0.15 s

5 0.20 s

6 0.25 s

7 0.30 s

How seatbelts work. Notice how the seatbelt stretches during the collision. This 'spreads' the change of the driver's momentum over a longer period, making the force he experiences smaller.

Questions

3 Using the diagrams on the right, estimate how long it takes for the seat belt to bring the driver's body to a stop. If his mass is 70 kg, what is the average force that the belt has to exert to do this?

4 A recent survey found that 95% of front-seat passengers wear seat belts. But only 69% of adult back-seat passengers wear a seat belt, though 96% of child back-seat passengers do.
   a Suggest why a few front-seat passengers do not wear a seat belt, despite the evidence that it reduces the risk of death.
   b Suggest why a higher proportion of children wear seatbelts.

Find out about

- ✓ **the laws (or rules) that apply to every example of motion**
- ✓ **how a resultant force is needed to change an object's motion**

Smooth floor

stops after a short distance

Imagine pushing this curling stone across a smooth floor. It will keep going after it leaves your hand, because of the momentum you have given it during the interaction with your hand. But it immediately begins to slow down because of friction, and soon it will stop.

Ice

goes further before stopping

Now think what would happen if you gave the same stone exactly the same push, but this time on ice. It would not slow down as quickly. But it would slow down eventually, because there is still some friction. Eventually it would stop.

'Perfect' ice

never stops!

Now imagine 'perfect' ice, so slippery that there is no friction force between it and the stone. If you were able to give the stone the same push as before, it would not slow down after it left your hand. There is no friction force, so it just keeps on going, at the same speed, for ever.

Steady speed

Already in this chapter, we've seen two 'rules' about forces and motion:
- if an object is stationary, the resultant force on it is zero
- if there is a resultant force acting on an object, it causes a change in the momentum of the object. If the object was initially at rest (stopped), a resultant force makes it accelerate (move with increasing speed).

What about an object moving at a steady speed? These rules do not say anything about this situation. What forces are involved? To answer this, think about the situation shown in the three diagrams.

So motion at a steady speed does not need a force to maintain it. If the resultant force on an object is zero, its motion will remain unchanged. If it happens to be stationary, it will stay stationary. If it happens to be moving, it will carry on moving at the same speed in the same direction.

In the real world

In the real world there is no 'perfect' ice as in the third diagram. So you can never have a moving object with no forces acting on it. But it is possible to have a situation where there is no *resultant* force on a moving object. This happens when the driving force that is causing the motion exactly balances the counter-forces (of friction and air resistance). When this occurs, the moving object does not slow down or stop. It keeps moving at whatever speed it had when the forces became equal.

To see how this works, think about the forces involved in riding a bicycle.

When you press on the pedals of a bike, the chain makes the back wheel turn. The tyre pushes back along the ground. The other force in this interaction pair is the force exerted by the ground on the tyre. This pushes the bike forward. It is therefore called the **driving force**. As you move, air resistance and friction at the axles cause a **counter-force**. This is in the opposite direction to your motion.

Slowing down

What happens if the cyclist stops pedalling? There is now no driving force, but there is still a counter-force. So the resultant force is now in the opposite direction to her direction of motion. It causes a change of momentum in the direction of the resultant force. This is a negative change, meaning that the cyclist's momentum gets smaller rather than bigger.

Some people find it hard to believe that an object can move in one direction, whilst the resultant force acting on it is in the other direction. But this happens all the time, in any situation where something has been set in motion and is now slowing down. The key is to remember that a force does not cause motion; it causes a *change* of motion.

Laws of motion

The laws of motion (worked out by Newton) can be stated as follows:

Law 1: If the resultant force acting on an object is zero, the momentum of the object does not change.

Law 2: If there is a resultant force acting on an object, the momentum of the object will change. The size of the change of momentum is equal to the resultant force × the time for which it acts. The change is in the same direction as the resultant force.

Law 3: When two objects interact, each experiences a force. The two forces are equal in size, but opposite in direction.

These laws apply to all objects and to every situation (apart from objects at the subatomic level or moving at speeds near the speed of light). They enable us to explain and predict the motion of objects.

 When you start off, the counter-force is small. Your driving force is bigger. You move forward, and your speed increases.

 As you go faster, the air resistance force on you gets bigger. The counter-force increases. You still get faster, but not as quickly as before.

 Eventually you reach a speed where the counter-force exactly balances your driving force. Your speed stops increasing. You are travelling, at a steady speed.

counter-force   driving force

Questions

1   List three examples from everyday life of a situation where the resultant force on an object is zero. Explain how these are in agreement with the first law of motion.

2   List three examples from everyday life of a situation where there is a resultant force acting on an object. Explain how these are in agreement with the second law of motion.

3   Draw a fourth diagram in the series on this page to show the forces acting when the cyclist stops pedalling and freewheels. Write a caption, like those for the first three diagrams, to explain the motion.

Key words
✓ **driving force**
✓ **counter-force**

Find out about

- ✔ **how to calculate the work done by a force**
- ✔ **the link between work done on an object and the energy transferred**
- ✔ **how to use energy ideas to predict the motion of objects**

Worked example

Calculate how much work is needed to push a car 50 m along a road.

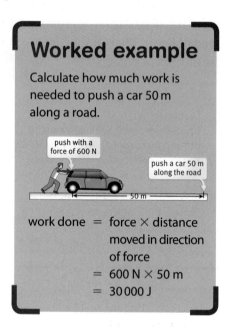

push with a force of 600 N

push a car 50 m along the road

50 m

$$\text{work done} = \text{force} \times \text{distance moved in direction of force}$$
$$= 600 \text{ N} \times 50 \text{ m}$$
$$= 30\,000 \text{ J}$$

Transferring energy

When you push something and start it moving, the force you exert gives the object momentum. But your push also transfers energy to the object. This comes from the energy stored in your body (which gets less) to the energy of the moving object (which gets bigger). The energy of a moving object is called **kinetic energy**.

In general, when a force makes something move, it transfers energy to the moving object or to its surroundings. We say that the force 'does **work**'. The amount of work it does depends on:

- the size of the force
- the distance the object moves in the direction of the force.

The equation for calculating the amount of work done by a force is:

$$\text{work done by a force} = \text{force} \times \text{distance moved in the direction of the force}$$

(joules, J) (newtons, N) (metres, m)

This also tells us the amount of energy transferred, because:

amount of energy transferred = amount of work done by the force

Energy and work are both measured in joules (J). A force of 1 newton applied over a distance of 1 metre does 1 J of work, and transfers 1 joule of energy.

Work

In physics, the word 'work' has a meaning that is slightly different from its everyday meaning.

Imagine that you are out in your car and you break down. Luckily, there is a garage just down the road. You ask your passenger to steer the car while you push it to the garage. If the garage is a long way down the road, you are going to have to do more work than if it is nearby.

However, if you push a really heavy object and cannot get it to move, you will feel that you are doing hard work that is having no result! From the physics point of view, you are not doing work because the force you are applying is not moving anything. The same is true if you are holding a heavy object. It feels like hard work, but the force you are applying to it is not making it move, so you are not doing any work.

Lifting things: gravitational potential energy

Lifting an object also involves doing work. Imagine lifting luggage into the car boot. You are transferring energy from your body's store of chemical energy. The **gravitational potential energy** of the luggage increases. The increase is equal to the amount of work you have done.

Suppose you have a suitcase that weighs 300 N. To lift it up, you have to exert an upward force of 300 N. If you lift it 1 metre into the boot of the car, then

work done = force × distance moved in the direction of the force

$$= 300 \text{ N} \times 1 \text{ m}$$

$$= 300 \text{ J}$$

The suitcase now has 300 J more gravitational potential energy. In general, when anything is lifted up, you can calculate the change in gravitational potential energy from the equation

gravitational potential energy = weight × vertical height difference
(joules, J)   (newtons, N)   (metres, m)

Notice that it is only the vertical height difference that matters. If you slide a suitcase up a ramp, the gain in gravitational potential energy is the same as if you lift it vertically.

Calculating kinetic energy

The equation for calculating the **kinetic energy** of a moving object is:

kinetic energy $= \dfrac{1}{2} \times$ mass $\times$ (velocity)^2
(joules, J)   (kilogram, kg)   (metres per second, m/s)^2

The kinetic energy is proportional to the mass of the moving object. But notice that the amount of kinetic energy depends on the velocity squared. So small changes in velocity mean quite big changes in kinetic energy.

I have to exert an upward lift force of just over 300 N to lift the suitcase.

1m

weight = 300 N

Doing work by lifting: increasing gravitational potential energy.

Questions

1 If your mass is 40 kg, then your weight is roughly 400 N. How much work do you have to do each time you go upstairs – a vertical height gain of 2.5 m?

2 A mother is pushing a child along in a buggy. She is doing work. So the amount of energy stored in her muscles is getting less. Where is this energy being transferred to? (Careful! The buggy is going at a steady speed.)

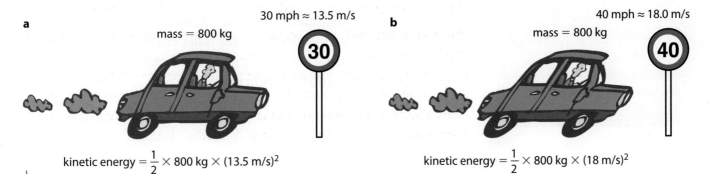

a

30 mph ≈ 13.5 m/s
mass = 800 kg

30

kinetic energy $= \dfrac{1}{2} \times 800$ kg $\times$ (13.5 m/s)^2

$= 72\,900$ J

b

40 mph ≈ 18.0 m/s
mass = 800 kg

40

kinetic energy $= \dfrac{1}{2} \times 800$ kg $\times$ (18 m/s)^2

$= 129\,600$ J

A car travelling at 40 mph (**b**) has nearly twice as much kinetic energy as the same car at 30 mph (**a**). It uses a lot more fuel to get to this speed.

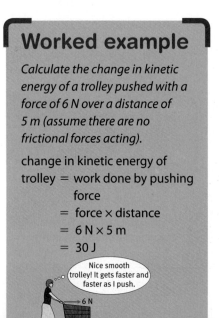

On a down slope, gravitational potential energy decreases and kinetic energy increases. On an up slope, the change is in the opposite direction.

Making things speed up: changing their kinetic energy

Imagine pushing a well-oiled supermarket trolley along a level floor. As you push, you are doing work. The trolley keeps speeding up. Its speed increases as long as you keep pushing. You are transferring energy from your body's store of chemical energy to the trolley, where it is stored as kinetic energy. If the trolley is absolutely smooth running, the amount of work you do pushing it is equal to the change in the trolley's kinetic energy.

A real trolley will always have some friction, so its change in kinetic energy will be less than this. Some work is wasted in causing unwanted heating.

Conservation of energy

Energy ideas are very useful for working out what will happen when something falls through a certain height difference, or slides down a smooth ramp. If friction is small enough to be ignored, the amount of gravitational potential energy stored in the system gets less, and the kinetic energy of the moving object gets bigger, by the same amount.

$$\text{decrease in gravitational potential energy} = \text{increase in kinetic energy}$$

The opposite happens if a moving object rises vertically, or up a smooth slope. Its kinetic energy decreases. The gravitational potential energy stored in the system gets bigger, by the same amount.

The really useful thing about this is that it doesn't depend on the shape of the path the object follows, as long as it is smooth enough to allow us to ignore friction. A fairground roller coaster is a good example of these ideas in action. As the roller coaster goes round the track, gravitational potential energy is changing to kinetic energy (on the down slopes) and vice versa (on the up slopes).

If the slope has a complicated shape, it would be very difficult, maybe even impossible, to work out how fast the roller coaster is going at the bottom of the slope using the ideas of force and momentum. It is easier to use the principle of **conservation of energy**.

To see how this works out in practice, look at the following example:

Worked example

Calculate the speed of the roller coaster shown below at the bottom of the ride (assuming there are no friction forces).

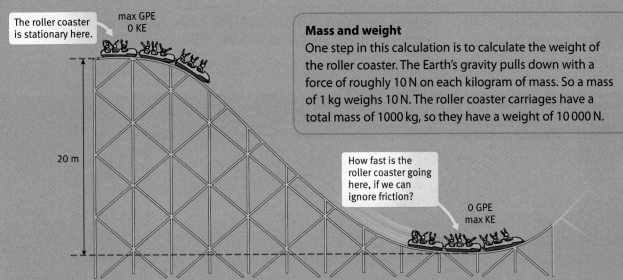

The roller coaster is stationary here.

max GPE
0 KE

Mass and weight
One step in this calculation is to calculate the weight of the roller coaster. The Earth's gravity pulls down with a force of roughly 10 N on each kilogram of mass. So a mass of 1 kg weighs 10 N. The roller coaster carriages have a total mass of 1000 kg, so they have a weight of 10 000 N.

20 m

How fast is the roller coaster going here, if we can ignore friction?

0 GPE
max KE

The simplest way to solve this problem is to use energy ideas. If friction is small enough to ignore, then the gravitational potential energy that the roller coaster loses as it goes down the slope is equal to the kinetic energy it gains:

$$\text{loss of gravitational potential energy} = \text{weight} \times \text{vertical height change}$$
$$= 10\,000\,\text{N} \times 20\,\text{m}$$
$$= 200\,000\,\text{J}$$

$$\text{loss of gravitational potential energy} = \text{gain in kinetic energy}$$

So $\text{gain in kinetic energy} = 200\,000\,\text{J}$

But $\text{gain in kinetic energy} = \frac{1}{2} \times \text{mass} \times (\text{velocity})^2$

So $\frac{1}{2} \times \text{mass} \times (\text{velocity})^2 = 200\,000\,\text{J}$

Multiply both sides by 2: $\text{mass} \times (\text{velocity})^2 = 400\,000\,\text{J}$

or $1000\,\text{kg} \times (\text{velocity})^2 = 400\,000\,\text{J}$

Divide both sides by 1000 kg: $(\text{velocity})^2 = \dfrac{400\,000\,\text{J}}{1000\,\text{kg}}$

$$= 400\,(\text{m/s})^2$$

Take the square root of both sides: velocity $= 20\,\text{m/s}$

The shape of the slope does not matter at all. Only the vertical height difference is important in working out the change in gravitational potential energy – and hence the increase in kinetic energy.

Questions

3 A ten-pin bowling ball has a mass of 4 kg. It is moving at 8 m/s. How much kinetic energy does it have?

4 Which of the following has more kinetic energy:
   a a car of mass 500 kg travelling at 20 m/s?
   b a car of mass 1000 kg travelling at 10 m/s?

5 Repeat the calculation above for a roller coaster that is only half as heavy (weight 5000 N). What do you notice about its speed at the bottom? How would you explain this?

Science
Explanations

Forces and motion form the basis of our understanding of how the world works. Every example of motion we observe can be explained by a few simple rules (or laws) that are so exact and precise that they can be used to predict the motion of an object very accurately.

You should know:

- about the interaction pair of forces that always arise when two objects interact
- that vehicles, and people, move by pushing back on something and this interaction causes a forward force to act on them
- about the friction interaction between two objects that are sliding (or tending to slide) past each other
- that the resultant force on an object is the sum of all the individual forces acting on it, taking their directions into account
- about the reaction force on an object that arises because it pushes down on a surface
- how to calculate the average speed of a moving object
- about instantaneous speed and how this is different to the average speed
- what is meant by distance, displacement, speed, velocity, and acceleration
- how to draw and interpret distance–time, speed–time, displacement–time, and velocity–time graphs
- how to calculate the acceleration of an object
- how to find acceleration from the gradient of a velocity–time graph
- that when a resultant force acts on an object it causes a change in momentum
- how to calculate momentum and the change in momentum due to a force
- that many vehicle safety features increase the duration of an event (such as a collision), so that the average force is less, for the same change of momentum
- that if there is no resultant force on an object, its momentum does not change – it either remains stationary or keeps moving at a steady speed in a straight line
- about the work done when a force moves an object and how to calculate the work done, which is equal to the energy transferred
- that when work is done on an object, energy is transferred to the object and when work is done by an object, energy is transferred from the object to something else
- about gravitational potential energy and how to calculate the change in gravitational potential energy as an object is raised or lowered
- about kinetic energy, how to calculate it, and that doing work on an object can increase its kinetic energy by making it move faster
- that when an object falls, if friction and air resistance can be ignored, the decrease in gravitational potential energy is equal to the increase in kinetic energy and this can be used to work out the speed of the object.

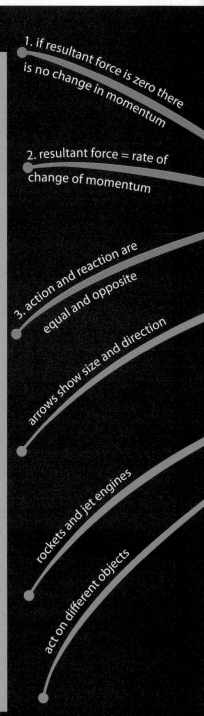

1. if resultant force is zero there is no change in momentum

2. resultant force = rate of change of momentum

3. action and reaction are equal and opposite

arrows show size and direction

rockets and jet engines

act on different objects

EXPLAINING MOTION

describing motion
- average speed and instantaneous speed
- distance, speed, acceleration, and time
- displacement, velocity, acceleration, time
- distance–time and displacement–time

plot and interpret graphs
- speed–time and velocity–time

mathematical skills

using equations
- displacement, velocity, acceleration, and time
- distance, speed, acceleration, and time
- mass, velocity, momentum, force, and time
- work, force, distance, kinetic energy, and gravitational potential energy
- resultant force, driving force, friction, reaction, counter force, air resistance

calculating with data
- mean
- range

- proportion
- correlation

developing explanations
- factors and outcomes
- data
- explanations
- predictions
  - accept
  - reject

the connection between force and motion
- Newton's laws of motion

momentum
- force and time
- change of momentum
- mass and velocity

forces
- reaction and friction forces
- interaction pairs
  - opposite direction
  - equal size

energy
- work done
  - force and distance
- kinetic energy
  - change in speed
- gravitational potential energy
  - change in height
- energy conservation

Ideas about Science

In addition to understanding forces and motion, you need to understand how scientific explanations are developed. The list below links these ideas about science with some examples from the module.

In developing scientific explanations you should be able to:

- identify statements that are data and statements that are explanations. For example, the statement 'The acceleration of a falling object has a constant value' is data; the statement 'When a constant force acts on an object it causes a constant acceleration' is an explanation.
- recognise that an explanation may be incorrect, even if the data is correct. For example, some measurements show that pushing with a constant force on a toy car causes it to travel at a constant speed, but the explanation 'A constant force causes a car to travel at a constant speed' is incorrect. The pushing force is balancing the friction force – there is no resultant force on the car.

- identify where creative thinking is involved in the development of an explanation. For example, the idea of an interaction pair of forces means that when you push on a wall the wall pushes back – this is not an obvious idea, but Isaac Newton suggested this as his third law of motion: action and reaction are equal and opposite.
- recognise data or observations that are accounted for, or conflict with, an explanation. For example, if data shows that the force on an object is proportional to its acceleration, this agrees with Newton's second law: the force on an object is equal to its rate of change of momentum.
- give good reasons for accepting or rejecting a scientific explanation. For example, Aristotle believed that heavier objects always fall faster than light objects. A good reason for rejecting this would be a slow-motion film of objects falling in a vacuum – they fall at the same speed.
- decide, giving reasons, which of two scientific explanations is better. Galileo said that light and heavy objects fall at the same speed. This matches observations better, when there is no air resistance.
- understand that when a prediction agrees with an observation this increases confidence in the explanation on which the prediction is based, but does not prove it is correct. For example, Newton's laws of motion have correctly predicted the behaviour of many moving objects.
- understand that when a prediction disagrees with an observation this indicates that one or the other is wrong and decreases confidence in the explanation on which the prediction is based. For example, observations do not support Aristotle's ideas about motion.

Review Questions

1 Think about the following situations:

i Amjad on his skateboard, throwing a heavy ball to his friend (main objects to consider: Amjad, the skateboard, and the ball)

ii a furniture remover trying to pull a piano across the floor, but it will not move (main objects to consider: the furniture remover, the piano, and the floor)

iii a hanging basket of flowers outside a café (main objects to consider: the basket and the chain it is hanging from).

For each situation:

a Sketch a diagram (looking at it from the side).

b Sketch separate diagrams of the main objects in the situation (these are listed for each).

c On these separate diagrams, draw arrows to show the forces acting on that object. Use the length of the arrow to show how big each force is.

d Write a label beside each arrow to show what the force is.

2 A tin of beans on a kitchen shelf is not falling, even though gravity is still acting on it. The shelf exerts an upward force, which balances the force of gravity. Explain in a short paragraph how it is possible for a shelf to exert a force. Draw a sketch diagram if it helps your explanation.

3 **a** The winner of a 50 m swimming event completes the distance in 80 s. What is his average speed?

b How far could Leonie cycle in 10 minutes if her average speed is 8 m/s?

c The average speed of a bus in city traffic is 5 m/s. How much time should the timetable allow for the bus to cover a 6-km route?

4 **a** A bus leaves a bus stop and reaches a speed of 15 m/s in 10 s. Calculate its acceleration.

b A car accelerates at 3 m/s^2 for 8 s. By how much will its speed have increased?

5 What is the momentum of:

a a hockey ball of mass 0.4 kg moving at 5 m/s?

b a jogger of mass 55 kg, running at 4 m/s?

c a van of mass 10 000 kg, travelling at 15 m/s?

d a car ferry of mass 20 000 000 kg, moving at 0.5 m/s?

6 A weightlifter raises a bar of mass 50 kg until it is above his head – a total height gain of 2.2 m. How much gravitational potential energy has it gained? How much more work must he do to hold it there for 5 s?

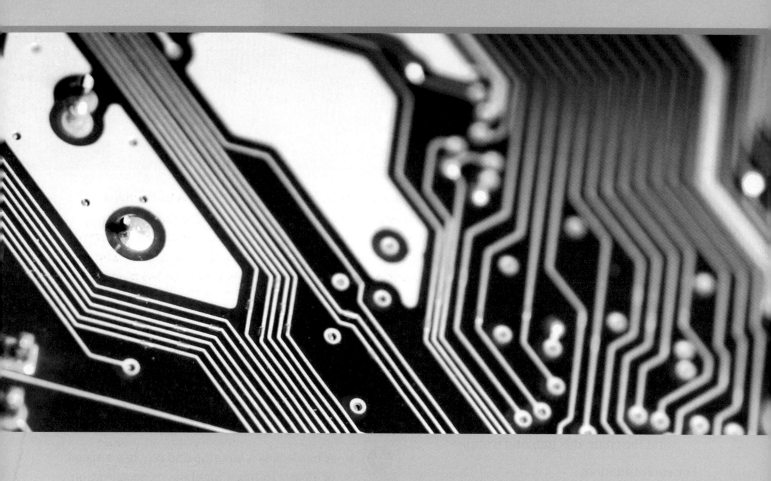

P5 Electric circuits

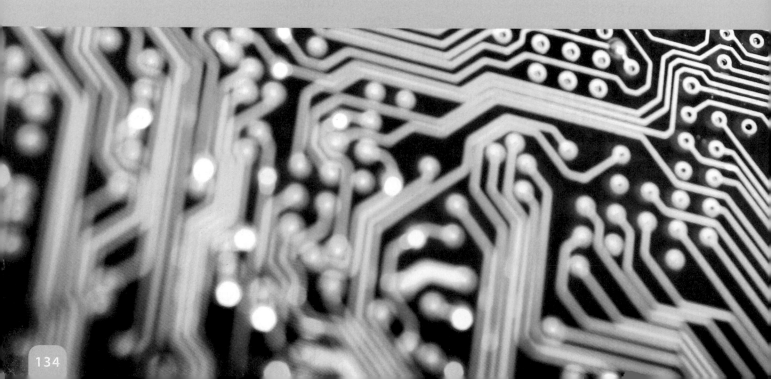

Why study electric circuits?

Imagine life without electricity – rooms lit by candles or oil lamps, no electric cookers or kettles, no radio, television, computers, or mobile phones, no cars or aeroplanes. Electricity has transformed our lives, but you need to know enough to use it safely. More fundamentally, electric charge is one of the basic properties of matter – so anyone who wants to understand the natural world around them needs to have some understanding of electricity.

What you already know

- Electric current is not used up in a circuit.
- Electric current transfers energy from the battery to components in the circuit.
- The power rating of a device is measured in watts and power = voltage × current.
- There is a magnetic field near a wire carrying a current. This can be used to make an electromagnet.
- In a power station a turbine drives a generator to produce electricity.

Find out about

- the idea of electric charge, and how moving charges result in an electric current
- how models that help us 'picture' what is going on in an electric circuit can be used to explain and predict circuit behaviour
- electric current, voltage, and resistance
- energy transfers in electric circuits, and how mains electricity is generated and distributed
- electric motors.

The Science

The particles that atoms are made of carry an electric charge. An electric current is a flow of charges. A useful model of an electric circuit is to imagine the wires full of charges, being made to move around by the battery. The size of current depends on the voltage and the resistance of the circuit. A voltage can be produced by moving a magnet near a coil.

Ideas about Science

Electrical properties of materials are used in many measuring instruments. To be sure that the reading gives a true value it is essential that the external conditions do not affect the reading.

Find out about

- electric charge and how it can be moved when two objects rub together
- the effects of like and unlike charges on each other
- the connection between charge and electric current

Why is it called 'charge'?

In the late 1700s, experimenters working on electricity thought that the effect you produced when you rubbed something was a bit like preparing a gun. In those days, you had to prime a gun by pushing down an explosive mixture into the barrel. This was called the 'charge'. When you had done this, the gun was 'charged'. It could be 'discharged' by firing.

Many of the properties of matter are obvious; matter has mass and it takes up space. But there is another property of matter that is less obvious. The most dramatic evidence of it is lightning. Other evidence comes from the shock you sometimes feel when you touch a car door handle after you get out. It is the property we call **electric charge**. Two hundred years ago, scientists were just beginning to understand electric charge, and to learn how to control it. Their work led to technological developments that have transformed our everyday lives.

Charging by rubbing

Electrical effects can be produced by rubbing two materials together. If you rub a balloon against your jumper, the balloon will stick to a wall. If you rub a plastic comb on your sleeve, the comb will pick up small pieces of tissue paper – they are attracted to the comb. In both cases, the effect wears off after a short time.

When you rub a piece of plastic, it is changed in some way; it can now affect objects nearby. The more it is rubbed, the stronger the effect. It seems that something is being stored on the plastic. If a lot is stored, it may escape by jumping to a nearby object, in the form of a spark. We say that the plastic has been charged. Charging by rubbing also explains the example discussed at the top of the page. When you get out of a car, you slide across the seat, rubbing your clothes against it.

Two types of charge

If you rub two identical plastic rods and then hold them close together, the rods push each other apart – they **repel**. The forces they exert on each other are very small, so you can only see the effect if one of the rods can move freely.

Lightning is the most striking and dramatic evidence of the electrical properties of matter. A lightning strike occurs when electric charges move at high speed from a thundercloud to the ground or vice versa.

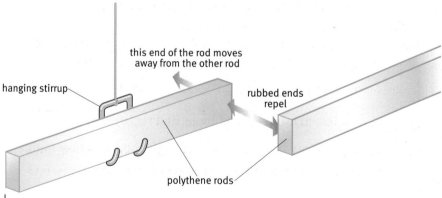

this end of the rod moves away from the other rod

hanging stirrup

rubbed ends repel

polythene rods

Two rubbed polythene rods repel each other. The hanging rod moves away from the other one.

If you try this with two rods of different plastics, however, you can find some pairs that **attract** each other. Scientists' explanation of this is that there are two types of electric charge. If two rods have the same type of charge, they repel each other. But if they have charges of different types, they attract. The early electrical experimenters called the two types of charge **positive** and **negative**. These names are just labels. They could have called them red and blue, or A and B.

Where does charge come from?

Scientists believe that charge is not *made* when two things are rubbed together it is just *moved around*. If you rub a plastic rod with a cloth, both the rod and the cloth become charged. (To see this, you need to wear a polythene glove on the hand holding the cloth, otherwise the charge will escape through your body.) Each object gets a different charge; if the rod has a positive charge, the cloth has a negative one. Rubbing does not make charge. It separates charges that were there all along.

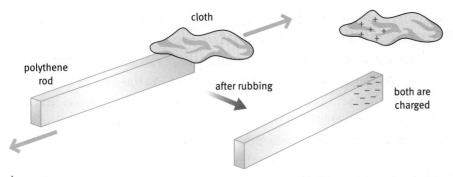

After it has been rubbed, the rod has a negative charge and the cloth has a positive charge. A possible explanation is that some electrons have been transferred from the cloth to the rod.

Questions

1 Imagine that you have two plastic rods that you know get a positive and a negative charge when you rub them with a cloth. Now you are given a third plastic rod. Explain how you could test whether it gets a positive or a negative charge when rubbed.

2 Some picture frames are made with plastic rather than glass. If you clean the plastic with a duster, it may get dusty again very quickly. Use the ideas on these pages to explain why this happens.

3 Use the ideas in the box on the right to explain why after you have rubbed a balloon on your jumper, the balloon will stick to the wall.

Attracting light objects

An object with a positive charge attracts another object with a negative charge. But why does a charged rod also attract light objects, such as little pieces of paper? The reason is that there are charges in the paper itself. Normally these are mixed up together, with equal amounts of each. So a piece of paper is uncharged. If a negatively charged rod comes near, it repels negative charges in the paper to the far end. This leaves a surplus of positive charges at the near end. The attraction between these positive charges and the rod is stronger than the repulsion between the negative charges (at the far end) and the rod. So the little piece of paper is attracted to the rod.

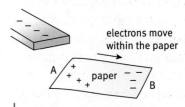

A charged rod separates the charges in a piece of paper nearby – and the paper is attracted to the rod.

Static electricity

The effects discussed here are often called electrostatic effects, and are said to be due to **static electricity**. The word 'static' indicates that the charges are fixed in position and do not move (except when there is a spark).

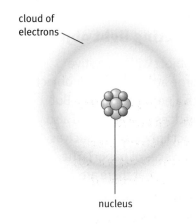

cloud of electrons

nucleus

What is electric charge?

Charge is a basic property of matter, which cannot be explained in terms of anything simpler. All matter is made of atoms, which in turn are made out of protons (positive charge), neutrons (no charge), and **electrons** (negative charge). In most materials there are equal numbers of positive and negative charges, so the whole thing is neutral. When you charge something, you move some electrons to it or from it.

The atom has a tiny nucleus made up of protons and neutrons. This has a positive charge. It is surrounded by a cloud of electrons, which have negative charge. As the electrons are on the outside of the atom they can be 'rubbed off', on to another object.

Although scientists cannot explain charge, they have developed useful ideas for predicting its effects. One is the idea of an **electric field**. Around every charge there is an electric field. In this region of space, the effects of the charge can be felt. Another charge entering the field will experience a force.

Moving charge = current

A van de Graaff generator is a machine for separating electric charge. When it is running, charge collects on its dome. As the charge builds up, the electric field around the dome gets stronger. The charge may 'jump' to another nearby object, in the form of a spark. If you hold a mains-testing screwdriver close to the dome and touch its metal end-cap with your finger the neon bulb inside the screwdriver lights up. There is an **electric current** through the lamp, making it light up. Charge on the dome is escaping across the air gap, through the neon bulb, and through you to the Earth. This (and other similar observations) suggests that an electric current is a flow of charge.

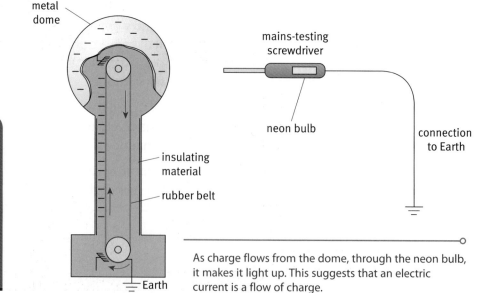

metal dome

mains-testing screwdriver

neon bulb

connection to Earth

insulating material

rubber belt

Earth

As charge flows from the dome, through the neon bulb, it makes it light up. This suggests that an electric current is a flow of charge.

A closed loop

The diagram on the right shows a simple **electric circuit**. If you make a circuit like this, you can quickly show that:

- if you make a break *anywhere* in the circuit, *everything* stops.

This suggests that something has to go all the way round an electric circuit to make it work. This 'something' is electric charge.

You will also notice that:

- both bulbs come on *immediately* when the circuit is completed. And they go off *immediately* if you make a break in the circuit.

This might seem to indicate that electric charge moves round the circuit very fast. Even with a large electric circuit and long wires, it is impossible to detect any delay. But there is another possible explanation. Perhaps there are **charges** (tiny particles with electric charge) in all the components of the circuit (wires, lamp filaments, batteries) *all the time*. Closing the switch allows these charges to move. They all move together, so the effect is immediate, even if the charges themselves do not move very fast.

The diagram below shows a model that is useful for thinking about how a simple electric circuit works. The power source is the hamster in the treadmill. As the treadmill turns, it pushes the peas along the pipe. If the pipe is full of peas all the time, then they will start moving everywhere around the circuit – immediately. The paddle wheel at the bottom will turn as soon as the hamster runs. The turning paddle wheel could be used to lift a mass. The hamster loses energy – and the mass gains energy. The hamster does work on the treadmill to make it turn and set the peas moving – and the moving peas do work on the paddle wheel.

Find out about

- ✔ how simple electric circuits work
- ✔ models that help explain and predict the behaviour of electric circuits
- ✔ how to measure electric current

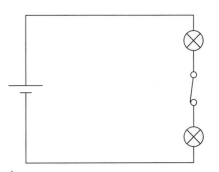

When you open the switch in this electric circuit, both bulbs go off.

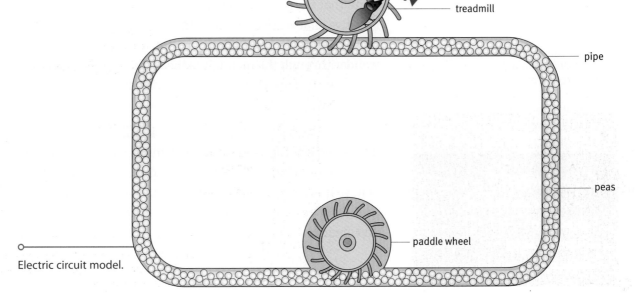

treadmill

pipe

peas

paddle wheel

Electric circuit model.

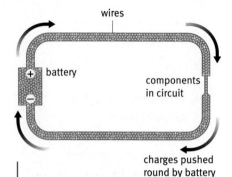

wires

battery

components
in circuit

charges pushed
round by battery

You can think of an electric current
as a flow of charges, which are
present in all materials (and free to
move in conductors), moving
round a closed conducting loop,
pushed by the battery.

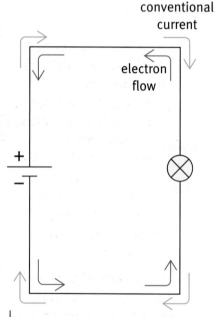

conventional
current

electron
flow

An electric current is a flow of
electrons through the wires of
the circuit. You can think of it
equally well as a 'conventional
current' of positive charges going
the other way.

Key words

✓ electric circuit
✓ series circuit
✓ ammeter
✓ charges

An electric circuit model

A model is a way of thinking about how something works. The
diagram on the right summarises a useful scientific model of an
electric circuit.

The key ideas are:
- charges are present throughout the circuit all the time
- when the circuit is a closed loop, the battery makes the charges
  move
- all of the charges move round together

The battery makes the charges move in the following way. Chemical
reactions inside the battery have the effect of separating electric
charges, so that positive charge collects on one terminal of the battery
and negative charge on the other. If the battery is connected to a
circuit, the charges on the battery terminals set up an electric field in
the wires of the circuit. This makes free charges in the wire drift
slowly along. However, even though the charges move slowly, they all
begin to move at once, as soon as the battery is connected. So the effect
of their motion is immediate. Notice too that the flow of charge is
continuous, all round the circuit. Charge also flows through the
battery itself.

Conventional current, electron flow

In the model above, the charges in the circuit are shown moving away
from the positive terminal of the battery, through the wires and other
components, and back to the negative terminal of the battery. This
assumes that the moving charges are positive. In fact there is no
simple way of telling whether the moving charges are positive or
negative, or which way they are moving. Long after this model was
first proposed and had become generally accepted, scientists came to
think that the moving charges in metals were electrons, which have
negative charge. In all metals, the atoms have some electrons that are
only loosely attached to their 'parent' atom and are relatively free to
wander through the metal. It is these that move, in the weak electric
field that the battery sets up in the wire. Insulators have few charges
free to flow.

To explain and predict how electric circuits behave, it makes no
difference whether you think in terms of a flow of electrons in one
direction or positive charges in the other. Although scientists now
believe it is electrons that flow in metals, in this course we use the
model of conventional current going the other way, as most physicists
and engineers do.

Electric current

An electric current is a flow of charge. You cannot see a current, but you can observe its effects. The current through a torch bulb makes the fine wire of the filament heat up and glow. The bigger the current through a bulb, the brighter it glows (unless the current gets too big and the bulb 'blows').

To measure the size of an electric current we use an **ammeter**. The reading (in amperes, or amps (A) for short) indicates the amount of charge going through the ammeter every second.

Current around a circuit

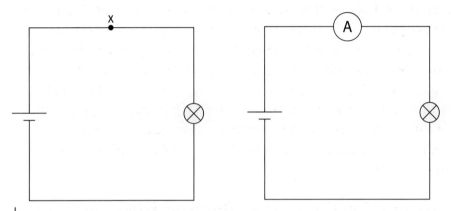

To measure the current at point X, you have to make a gap in the circuit at X and insert the ammeter in the gap, so that the current flows through it.

If you use an ammeter to measure the size of the electric current at different points around a circuit, you get a very important result.

* The current is the same everywhere in a simple (single-loop) electric circuit.

This may seem surprising. Surely the bulbs must use up current to light. But this is not the case. Current is the movement of charges in the wire, all moving round together like dried peas in a tube. No charges are used up. The current at every point round the circuit must be the same.

Of course, *something* is being used up. It is the energy stored in the battery. This is getting less all the time. The battery is doing work to push the current through the filaments of the light bulbs, and this heats them up. The light then carries energy away from the glowing filament. So the circuit constantly transfers energy from the battery, to the bulb filaments, and then on to the surroundings (as light). The current enables this energy transfer to happen. But the current itself is not used up.

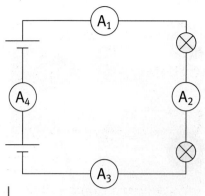

The current is the same size at all these points – even between the batteries. Current is not used up to make the bulbs light. This is a **series** circuit.

Questions

1 Look at the electric circuit model on page 139. What corresponds to:
   i the battery, ii the electric current, iii the size of the electric current?
   a Suggest one thing in a real electric circuit that might correspond to the paddle wheel.
   b How might you model a switch?

2 How would you change the electric circuit model to explore what happens in a series circuit with two identical bulbs? Use the model to explain:
   a why both bulbs go on and off together when the circuit is switched on and off
   b why both bulbs light immediately when the circuit is switched on
   c why both bulbs are equally bright.

Often, we want to run two or more things from the same battery. One way to do this is to put them all in a single loop, one after the other. Components connected like this are said to be **in series**. All of the moving charges then have to pass through each of them.

Another way is to connect components **in parallel**. In the circuit on the left, the two bulbs are connected in parallel. This has the advantage that each bulb now works independently of the other. If one burns out, the other will stay lit. This makes it easy to spot a broken one and replace it.

Currents in the branches

Look at the circuit below, which has a motor and a buzzer connected in parallel. A student, Nicola, was asked to measure the current at points a, b, c, and d. Her results are shown in the table below on the right.

In this portable MP3 player, the battery has to run the motor that turns the hard drive, the head that reads the disk, and the circuits that decode and amplify the signals.

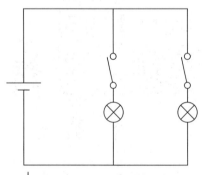

An advantage of connecting components in parallel is that each can be switched on and off independently.

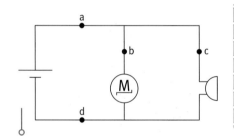

Point in circuit	Current (mA)
a	230
b	150
c	80
d	230

Measuring currents in a circuit with two parallel branches.

Nicola noticed that the current is the same size at points a and d: 230mA. When she added the currents at b and c, the result was also 230mA. This makes good sense if you think about the model of charges moving round. At the junctions, the current splits, with some charges flowing through one branch and the rest flowing through the other branch. Current is the amount of charge passing a point every second. So the amounts in the two branches must add up to equal the total amount in the single wire before or after the branching point.

Key words
- ✓ **in series**
- ✓ **in parallel**

Questions

1 When we want to run several things from the same battery, it is much more common to connect them in parallel than in series. Write down three advantages of parallel connections.

2 Look at the circuit above on the left. If you wanted to switch both bulbs on and off together, where would you put the switch? Draw two diagrams showing two possible positions of the switch that would do this.

3 In the circuit above with the motor and buzzer, what size is the electric current:
   a in the wire just below the motor?
   b in the wire just below the buzzer?
   c through the battery itself?

Controlling the current

The scientific model of an electric circuit imagines a flow of charge round a closed conducting loop, pushed by a battery. This movement of charge all round the circuit is an electric current.

The size of the current is determined by two factors:

- the **voltage** of the battery
- the **resistance** of the circuit components.

Battery voltage

Batteries come in different shapes and sizes. They usually have a voltage measured in volts (V), marked clearly on them, for example, 1.5 V, 4.5 V, 9 V. To understand what voltage means and what this number tells you, look at the following diagrams, which show the same bulb connected first to a 4.5 V battery and then to a 1.5 V battery. Voltage is measured using a voltmeter.

Find out about

✔ **how the battery voltage and the circuit resistance together control the size of the current**
✔ **what causes resistance**
✔ **the links between battery voltage, resistance, and current**

All the batteries on the front row are marked 1.5 V – but are very different sizes. The three at the back are marked 4.5 V, 6 V, and 9 V.

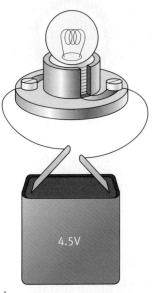

With a 4.5 V battery, this bulb is brightly lit.

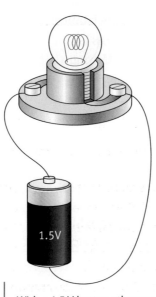

With a 1.5 V battery, the same bulb is lit, but very dimly.

The bigger the current through a light bulb, the brighter it will be (up to the point where it 'blows'). So the current through the bulb above is bigger with the 4.5 V battery. You can think of the voltage of a battery as a measure of the 'push' it exerts on the charges in the circuit. The bigger the voltage, the bigger the 'push' – and the bigger the current as a result.

The battery voltage depends on the choice of chemicals inside it. To make a simple battery, all you need are two pieces of different metal and a beaker of salt solution or acid. The voltage quickly drops, however. The chemicals used in real batteries are chosen to provide a steady voltage for several hours of use.

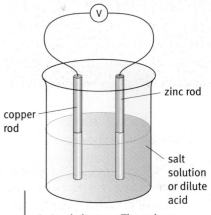

A simple battery. The voltage depends on the metals and the solution you choose.

Resistance

The size of the current in a circuit also depends on the resistance that the components in the circuit provide to the flow of charge. The battery 'pushes' against this resistance. You can see the effect of this if you compare two circuits with different resistors. Resistors are components designed to control the flow of charge.

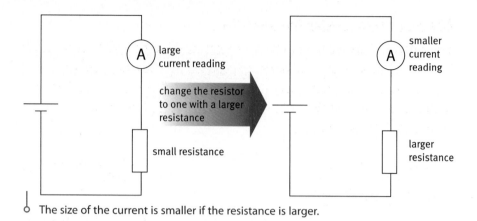

The size of the current is smaller if the resistance is larger.

Changing resistance changes the size of the current. The bigger the resistance, the smaller the current.

What causes resistance?

Everything has resistance, not just special components called resistors. The resistance of connecting wires is very small, but not zero. Other kinds of metal wire have larger resistance. The filament of a light bulb has a lot of resistance. This is why it gets so hot and glows when there is a current through it. A heating element, like that in an electric kettle, is a resistor.

All metals get hot when charge flows through them. In metals, the moving charges are free electrons. As they move round, they collide with the fixed array, or lattice, of ions in the wire. These collisions make the ions vibrate a little more, so the temperature of the wire rises. In some metals, the ions provide only small targets for the electrons, which can get past them relatively easily. In other metals, the ions present a much bigger obstacle – and so the resistance is bigger.

Why the temperature of a wire rises when a current flows through it.

Questions

1 Look back at the electric circuit model in Section B. How would you change the model to show the effect of:
   a a bigger battery voltage?
   b an increase in resistance?
   (You might be able to think of two different ways of doing this.)
   Does the model correctly predict the effect these would have on the current?

2 Suggest two different ways in which you could change a simple electric circuit to make the electric current bigger.

Key relationships in an electric circuit: A summary

The size of the electric current (*I*) in a circuit depends on the battery voltage (*V*) and the resistance (*R*) of the circuit.

• If you make *V* bigger, the current (*I*) increases.

• If you make *R* bigger, the current (*I*) decreases.

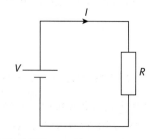

Measuring resistance

The relationship between battery voltage, resistance, and current leads to a way of measuring and defining resistance. The first step is to explore the relationship between voltage and current in more detail. Keiko, a student, did this by measuring the current through a coil of wire with different batteries.

1 Keiko connected a coil of resistance wire to a 1.5 V battery and an ammeter. She noted the current.

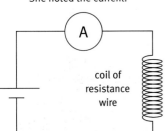

coil of resistance wire

2 She then added a second battery in series. She noted the current again.

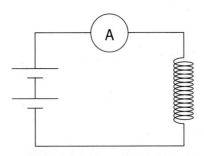

3 Keiko repeated this with 3, 4, 5, and 6 batteries, to get a set of results:

4 Finally, she drew a graph of current against battery voltage:

Number of 1.5 V batteries	Battery voltage (V)	Current (mA)
1	1.5	75
2	3.0	150
3	4.5	225
4	6.0	300
5	7.5	375
6	9.0	450

The straight-line graph means that the current in the circuit is proportional to the battery voltage. This result is known as **Ohm's law**. The number you get if you divide voltage by current is the same every time. The bigger the number, the larger the resistance. This is how to measure resistance:

$$\text{resistance of a conductor} = \frac{\text{voltage across the conductor}}{\text{current through the conductor}}$$

$$R = \frac{V}{I}$$

The units of resistance are called ohms (Ω).

Rearranging this equation gives $I = V/R$. You can use this to calculate the current in a circuit, if you know the battery voltage and the resistance of the circuit.

Questions

3 In Keiko's investigation:
  a how many 1.5 V batteries would she need to use to make a current of 600 mA flow through her coil?
  b what is the resistance of her coil, in ohms?

4 In a simple series circuit, a 9 V battery is connected to a 45 Ω resistor.

  What size is the electric current in the circuit?

Ohm's law

Ohm's law says that the current through a conductor is **proportional** to the voltage across it – provided its temperature is constant. It only applies to some types of conductor (such as metals). An electric current itself causes heating, which complicates matters. For example, the current through a light bulb is not proportional to the battery voltage. The *I–V* graph is curved. The reason is that bulb filament heats up – and its resistance increases with temperature.

Key words
- ✔ voltage
- ✔ resistance
- ✔ Ohm's law

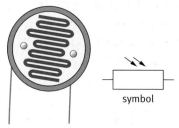

Using a variable resistor to control the current in a series circuit.

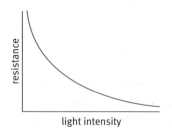

The resistance of a light-dependent resistor decreases as the light intensity increases.

symbol

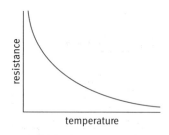

The resistance of a thermistor decreases as the temperature increases.

symbol

Variable resistors

Resistors are used in electric circuits to control the size of the current. Sometimes we want to vary the current easily, for example, to change the volume on a radio or CD player. A variable resistor is used. Its resistance can be steadily changed by turning a dial or moving a slider.

The circuit diagram on the left shows the symbol for a variable resistor. As you alter its resistance, the brightness of the light bulb changes, and the readings on *both* ammeters increase and decrease together. The variable resistor controls the size of the current everywhere round the circuit loop. (Is this what the circuit model in Section B would predict?)

Each of these sliders adjusts the value of a variable resistor.

Some useful sensing devices are really variable resistors. For example, a light-dependent resistor (LDR) is a semiconductor device whose resistance is large in the dark but gets smaller as the light falling on it gets brighter. An LDR can be used to measure the intensity (brightness) of light or to switch another device on and off when the intensity of the light changes. For example, it could be used to switch an outdoor light on in the evening and off again in the morning.

A thermistor is another device made from semiconducting material. Its resistance changes rapidly with temperature. The commonest type has a lower resistance when it is hotter. Thermistors can be used to make thermometers (to measure temperature) or to switch a device on or off as temperature changes. For example, a thermistor could be used to switch an immersion heater on when the temperature of water in a tank falls below a certain value and off again when the water is back at the required temperature.

Combinations of resistors

Most electric circuits are more complicated than the ones discussed so far. Circuits usually contain many components, connected in different ways. There are just two ways of connecting circuit components: in series or in parallel.

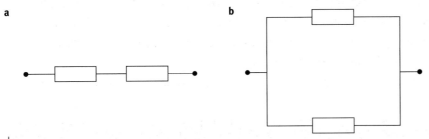

○ Two resistors connected **a** in series, **b** in parallel.

Two resistors in series have a larger resistance than one on its own. The battery has to push the charges through both of them. But connecting two resistors in parallel makes a smaller total resistance. There are now two paths that the moving charges can follow. Adding a second resistor in parallel does not affect the original path but adds a second equivalent one. It is now easier for the battery to push charges round, so the resistance is less.

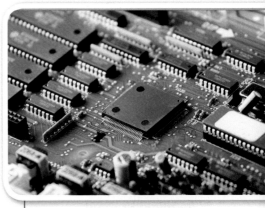

This circuit board from a computer contains a complex circuit, with many components.

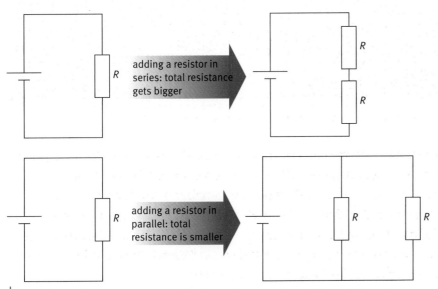
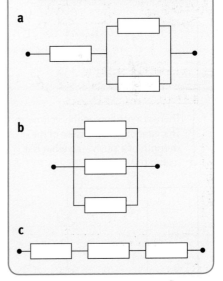

○ Different ways of adding a second resistor: how you do it makes a difference.

Question

5 All the resistors in the three diagrams below are identical. Put the groups of resistors in order, from the one with the largest total resistance to the one with the smallest total resistance.

a

b

c

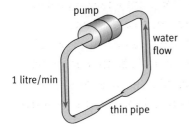
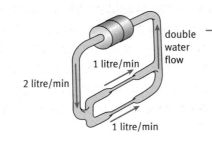

A water-flow model shows how the total resistance gets less when a second parallel path is added.

You can think of the voltage of a battery as a measure of the 'push' it exerts on the charges in a circuit. But a voltmeter also shows a reading if you connect it across a resistor or bulb in a working circuit as in the circuit on the left. Resistors and bulbs do not 'push'. So the **voltmeter** reading must be indicating something else.

A useful picture is to think of the battery as a pump, lifting water up to a higher level. The water then drops back to its original level as it flows back to the inlet of the pump. The diagram below shows how this would work for a series circuit with three resistors (or three lamps). The pump does **work** on the water to raise it to a higher level. This increases the gravitational potential energy stored. The water then does work on the three water wheels as it falls back to its original level. If we ignore energy losses, this is the same as the amount of work done by the pump.

In the electric circuit, the battery does work on the electric charges, to lift them up to a higher 'energy level'. They then do work (and transfer energy) in three stages as they drop back to their starting level. A voltmeter measures the difference in 'level' between the two points it is connected to. This is called the potential difference between these points. **Potential difference (p.d.)** is measured in volts (V).

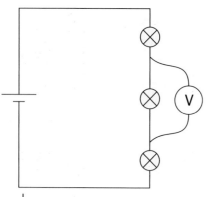

There is a reading on this voltmeter. This cannot be a measure of the strength of a 'push' – so what is it telling us?

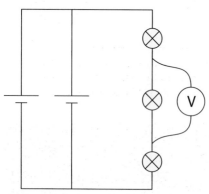

Adding a second battery in parallel does not change the potential difference across the lamps. So the current through the lamps does not change.

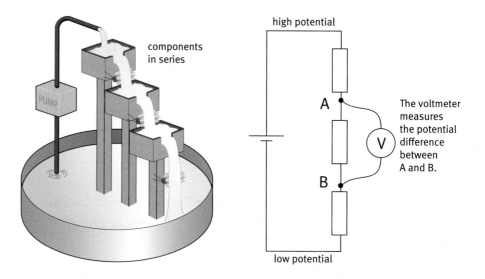

components in series

high potential

A

B

low potential

The voltmeter measures the potential difference between A and B.

The voltage of a battery is the potential difference between its terminals. If you put a battery with a larger voltage into the circuit above, this would mean a bigger potential difference across its terminals. The potential difference across each lamp (or resistor) would also now be bigger. Going back to the water-pump model, this is like changing to a stronger pump that lifts the water up to a higher level. The three downhill steps then also have to be bigger, so that the water ends up back at its starting level.

The same idea works for a parallel circuit. In this case, the water divides into three streams. Each loses all its energy in a single step.

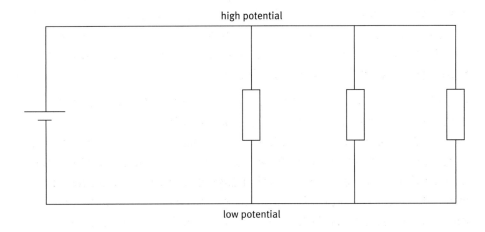

high potential

low potential

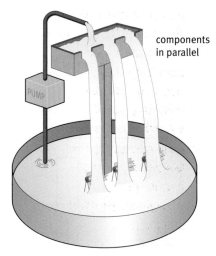

components in parallel

Voltmeter readings across circuit components

This water-pump model helps to explain and to predict voltmeter readings across resistors in different circuits. If several resistors are connected in parallel to a battery, the potential difference across each is the same. It is equal to the battery voltage.

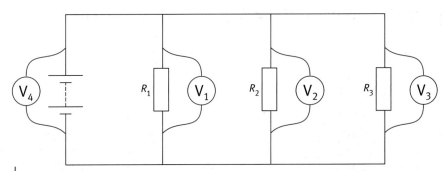

All of these voltmeters will have the same reading, even if R_1, R_2, and R_3 are different.

If the resistors are connected in series (as on the right), the sum of the potential differences across them is equal to the battery voltage. This is exactly what you would expect from the 'waterfall' picture on the previous page.

In the series circuit, the potential difference across each resistor depends on its resistance. The biggest voltmeter reading is across the resistor with biggest resistance. Again this makes sense. More work has to be done to push charge through a big resistance than a smaller one.

If several identical batteries are connected in parallel with a single resistor, the potential difference across the resistor remains the same as for one battery. As the potential difference across the resistor does not change, the current through the resistor does not change.

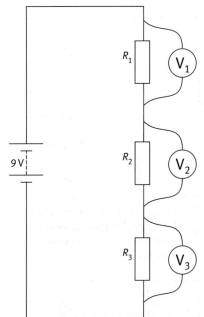

The voltages are in proportion to the resistances, and their sum is equal to the battery voltage.

Key words

- ✔ **voltmeter**
- ✔ **potential difference (p.d.)**
- ✔ **work**

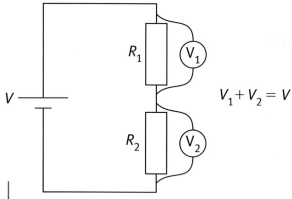

$$V_1 + V_2 = V$$

Two resistors in series make a potential divider.

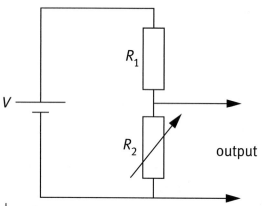

output

Making a variable voltage supply.

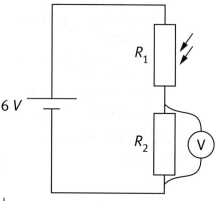

A simple light intensity meter.

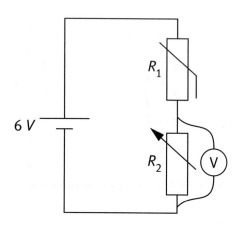

Potential divider

A circuit with two resistors in series has many important applications, particularly in electronics. The sum of the potential differences across the two resistors is equal to the p.d. across the battery (the battery voltage). The resistors divide up the battery voltage into two parts. For this reason, this kind of circuit is often called a **potential divider**. The two parts are often unequal. The bigger p.d. is always across the bigger resistance.

One application is to make a variable voltage supply. If the resistance of the variable resistor R_1 is turned right down to zero, then the potential difference across the output leads will be zero. The p.d. across the fixed resistor R_2 will be equal to the battery voltage. But if the resistance of R_1 is now increased, a larger share of the total will appear across it. The potential difference across the output leads will increase. Unlike a battery with a fixed voltage, we have a power supply with a voltage that we can change by turning a control knob.

Sensors also use the potential divider principle. A light sensor uses a light dependent resistor (LDR) in series with a fixed resistor. In the dark, the resistance of the LDR is large – much bigger than the resistance of R_2. So the reading on the voltmeter is very small, close to zero.

But in bright light, the resistance of the LDR falls dramatically until it is very small. Now the resistance of R_2 is much bigger than the resistance of the LDR. So the p.d. across R_2 is almost all of the battery voltage – and the reading on the voltmeter is close to 6 V.

So the reading on the voltmeter indicates the brightness of the light falling on the LDR.

Question

1 Describe and explain the reading on the voltmeter in this circuit when the thermistor is:

a cold

b hot.

Explain how the circuit could be used as an electrical thermometer.

Currents in parallel branches

The potential difference across resistors R_1, R_2, and R_3 in the parallel circuit earlier is exactly the same for each. It is equal to the p.d. across the battery itself. But the *currents* through the resistors are not necessarily the same. This will depend on their resistances. The current through the biggest resistor will be the smallest. There are two ways to think of this.

1   Imagine water flowing through a large pipe. The pipe then splits in two, before joining up again later. If the two parallel pipes have different diameters, more water will flow every second through the pipe with the larger diameter. The wider pipe has less resistance than the narrower pipe to the flow of water. So the current through it is larger.

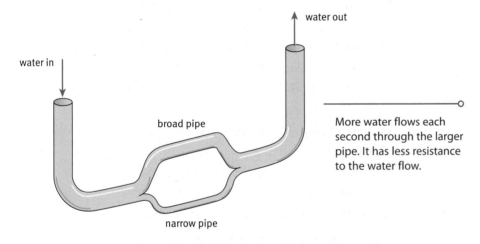

water out

water in

broad pipe

narrow pipe

More water flows each second through the larger pipe. It has less resistance to the water flow.

2   Think of two resistors connected in parallel to a battery as making two separate simple-loop circuits that share the same battery. The current in each loop is independent of the other. The smaller the resistance in a loop, the bigger the current. Some wires in the circuit are part of both loops, so here the current will be biggest. The current here will be the sum of the currents in the loops.

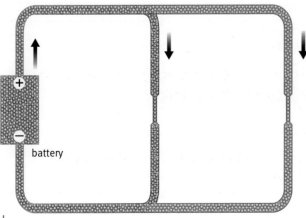

battery

A parallel circuit like this, behaves like two separate simple-loop circuits.

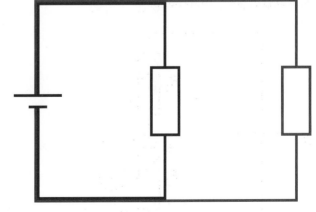

> **Question**
>
> 2   Imagine removing the red resistor from the circuit below leaving a gap. What would happen to:
> a   the current through the purple resistor?
> b   the current from (and back to) the battery?
>
> Explain your reasoning each time.

Find out about

- ✓ **how the power produced in a circuit component depends on both current and voltage**

An electric circuit is primarily a device for doing work of some kind. It transfers energy initially stored in the battery to somewhere else. A key feature of any electric circuit is the rate at which work is done on the components in the circuit – that is, the rate at which energy is transferred from the battery to the other components and on into the environment. This is called the **power** of the circuit.

Measuring the power of an electric circuit

Imagine starting with a simple battery and bulb circuit and trying to double, and treble, the power. You could do this in two ways.

- One is to add a second bulb, and then a third, in parallel with the first. In the circuits down the left-hand side of the diagram below, the p.d. is the same, but the current supplied by the battery doubles and trebles. The power is proportional to the current.
- Another is to add a second bulb, and then a third, in series with the first. Now you need to add a second battery, and then a third, to keep the brightness of the bulbs the same each time. In these circuits (across the diagram below), the current is the same, but the p.d. of the battery doubles and trebles. The power is proportional to the voltage.

This is summarised in the box in the bottom right-hand corner.

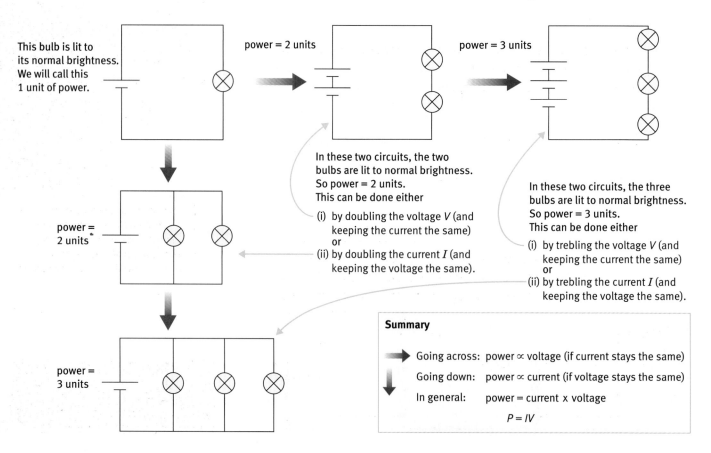

This bulb is lit to its normal brightness. We will call this 1 unit of power.

power = 2 units

power = 3 units

power = 2 units*

power = 3 units

In these two circuits, the two bulbs are lit to normal brightness. So power = 2 units. This can be done either

(i) by doubling the voltage *V* (and keeping the current the same)
or
(ii) by doubling the current *I* (and keeping the voltage the same).

In these two circuits, the three bulbs are lit to normal brightness. So power = 3 units. This can be done either

(i) by trebling the voltage *V* (and keeping the current the same)
or
(ii) by trebling the current *I* (and keeping the voltage the same).

Summary

➤ Going across: power ∝ voltage (if current stays the same)

⬇ Going down: power ∝ current (if voltage stays the same)

In general: power = current × voltage

$$P = IV$$

In general, the power dissipated in an electric circuit depends on both the current and the voltage:

power	=	current	×	voltage
P	=	I		V
(watt, W)		(ampere, A)		(volt, V)

The unit of power is the watt (W). One watt is equal to one joule per second.

To see how this equation for power makes sense, look back at the explanation of resistance and heating on page 144. If the battery voltage is increased, the electric field in the wires gets bigger. So the free charges (the electrons) move faster. When a charge collides with an atom in the wire, more energy is transferred in the collision. These collisions also happen more often, simply because the charges are moving along faster. So when the voltage is increased, collisions between electrons and the lattice of atoms are *both* harder *and* more frequent.

If you know the power, it is easy to calculate how much work is done (or how much energy is transferred) in a given period of time:

work done (or energy transferred)	=	power	×	time
(joule, J)		(watt, W)		(second, s)

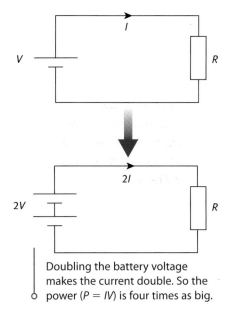

Doubling the battery voltage makes the current double. So the power ($P = IV$) is four times as big.

The power of the electric motor in this tube train is much greater than the power of the strip light above the platform. Both the voltage and the current are bigger.

Questions

1 In these two circuits, resistor R_1 has a large resistance and resistor R_2 has a small resistance. If each circuit is switched on for a while, which resistor will get hotter? Explain your answers.

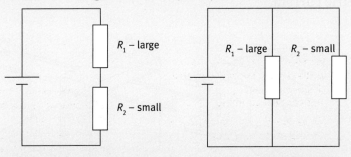

2 In circuit A, a battery is connected to a resistor with a small resistance. In circuit B, the resistor has a large resistance. The two batteries are identical. Which will go 'flat' first? Explain your answer.

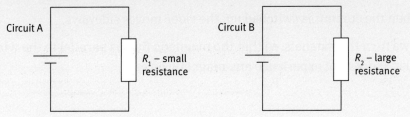

Key word

✓ **power**

Find out about

- ✓ **the force on a current-carrying wire in a magnetic field**
- ✓ **why an electric motor spins**

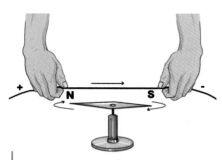

The wire is above the compass needle. The needle moves when the electric current is switched on.

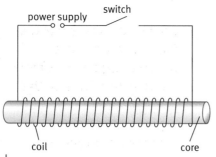

A coil of wire wound round an iron core makes an electromagnet – a magnet that can be switched on and off.

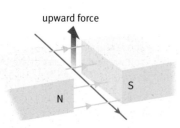

The magnetic force on the wire is at right angles to both the magnetic field lines and the electric current.

Magnetic effects

In 1819, the Danish physicist Hans-Christian Oersted noticed that the needle of a magnetic compass moved every time he switched on an electric current in a nearby wire. He investigated this further, and showed that there was a link between electricity and magnetism. When there is an electric current in a wire, there is a **magnetic field** in the region around the wire. The compass needle is a magnet and experiences a force because it is in the magnetic field caused by the electric current.

Winding a wire into a coil makes the magnetic field stronger. This is because the fields of each turn of the coil add together. It can be strengthened further by putting an iron core inside the coil to make an **electromagnet**.

Magnetic forces

A permanent magnet, such as a compass needle, experiences a force if it is placed in the magnetic field near a wire that is carrying an electric current. What if we keep the magnet fixed and allow the wire to move? The diagram shows one way of doing this. The two long parallel wires, and the wire 'rider' that is laid across them, are all made of copper wire with no insulation. The 'rider' is sitting in the magnetic field between the two flat magnets on the metal holder. If you switch on the power supply, the 'rider' slides sideways. The force acting on it is at right angles to both the magnetic field lines and to the electric current.

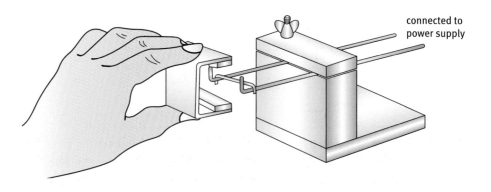

connected to power supply

When the current is switched on, the rider moves sideways.

If we turn the magnets, so that the magnetic field is parallel to the wire rider, it does not experience any magnetic force.

Turning effect on a coil

Magnetic forces can make a wire coil turn when a current flows. The diagram below shows a square coil of wire in a magnetic field. There will be no forces on the two ends of the coil, because the currents in these wires are parallel to the magnetic field lines. There will, however, be forces on the two sides of the coil, because the currents in these are at right angles to the magnetic field lines. The forces will be at right angles to both the field and the current. One force is up and the other down, because the currents in the two sides of the coil are in opposite directions. The effect of these forces is to make the coil rotate around the dotted line. If the coil is made with several turns of wire, this will make the turning forces stronger.

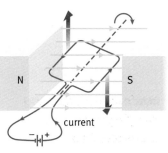

Turning effect of the magnetic forces on a flat coil

An electric motor

The coil in the diagram above would only turn through 90° before stopping. But if we could reverse the direction of the current in the coil at this point, this would reverse the direction of the forces on each side – and keep it turning for a further half-turn. If we could then reverse the current direction again, we could make the coil turn continuously. This is how a simple electric **motor** works.

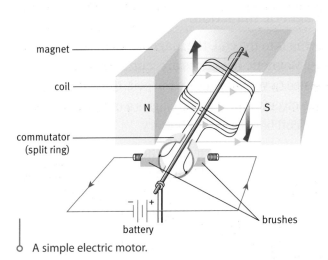

A simple electric motor.

Rather than having fixed wires, the electric circuit now includes a pair of brushes that rub against a split-ring **commutator**. This is fixed to the coil and rotates with it. As the coil rotates, each half of the split ring touches one brush for half a turn, and the other brush for the next half-turn. So the current direction in the coil is reversed twice every turn – changing the direction of the magnetic forces, and keeping the coil rotating.

Questions

1 Look at the diagram of the flat coil between the magnets. Explain why there is no rotating force on the coil when it is vertical.
2 Explain how the commutator ensures that the current in the coil changes direction even though it is connected to a battery with a direct current.

Key words

✔ **magnetic field**
✔ **electromagnet**
✔ **motor**
✔ **commutator**

Find out about

- ✔ **how a magnet moving near a coil can generate an electric current**
- ✔ **the factors that determine the size of this current**
- ✔ **how this is used to generate electricity on the large scale**

There is a connection between electricity and magnetism. An electric current generates a magnetic field, which can then be used to cause motion, as in an electric motor. But can we do this in reverse? Is it possible to generate an electric current by moving a wire near a magnet?

Electromagnetic induction

In the 1830s, the English physicist Michael Faraday did a series of investigations with magnets, wires, and coils. He found that he could generate an electric current by moving a magnet into a coil of wire. There was a current only while the magnet was moving, not while it was stationary inside the coil. When he pulled the magnet out of the coil again, there was another current but now in the opposite direction. This effect is called **electromagnetic induction**.

Michael Faraday discovered electromagnetic induction.

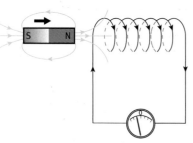

1. While the bar magnet is moving into the coil, there is a small reading on the sensitive ammeter.

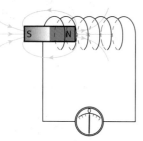

2. There is no current while the magnet is stationary inside the coil.

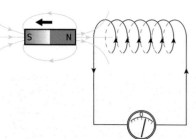

3. While the magnet is being removed from the coil, there is again a small current, but now in the opposite direction.

Moving a magnet into, or out of, a coil generates a current.

While the magnet is moving into the coil, the magnetic field around the wires of the coil changes. Magnetic field lines are 'cutting' the coil. This causes a potential difference (a voltage) across the coil. If the coil is connected into a complete circuit, this voltage causes a current. While the magnet is moving, the coil behaves like a battery.

There is no voltage (and hence no current) while the magnet is stationary inside the coil. It is *changes* in the magnetic field, not the field itself, that cause the induced voltage. When the magnet is pulled out again, the magnetic field changes again. There is an induced voltage (and current) in the other direction.

The size of the induced voltage can be increased by:
- moving the magnet in and out more quickly
- using a stronger magnet
- using a coil with more turns (there is an induced voltage in each turn of the coil, and these add together).

Making a generator

We can make a simple **generator** by rotating a magnet near one end of a coil. The effect is more noticeable if we put an iron core into the coil. This greatly increases the strength of the magnetic field inside the coil. As the magnet rotates, the magnetic field around the coil is constantly changing. This induces a voltage across the ends of the coil, which

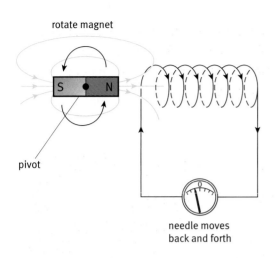

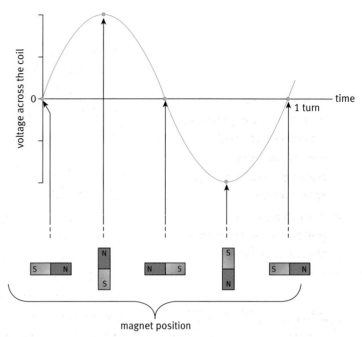

When the magnet rotates, it induces an alternating voltage across the coil.

causes an electric current in the circuit. The voltage and current change direction every half-turn of the magnet. This arrangement generates an **alternating current (a.c.)** in the circuit. For many applications, a.c. works just as well as the **direct current (d.c.)** produced by a battery. A direct current is one that flows in just one direction.

The size of the alternating voltage produced by a generator of this sort can be increased by:

*   using a stronger rotating magnet or electromagnet
*   rotating the magnet or electromagnet faster (though this also affects the frequency of the a.c. produced)
*   using a fixed coil with more turns
*   putting an iron core inside the fixed coil (this makes the magnetic field a lot bigger – as much as 1000 times).

In a typical power station generator, an electromagnet is rotated inside a fixed coil. As it spins, a.c. is generated in the coil. In power stations in the UK, the rate of turning is set at 50 cycles per second. The generator is turned by a turbine, which is driven by steam. The steam is produced by burning gas, oil, coal, or by the heating effect of a nuclear reaction.

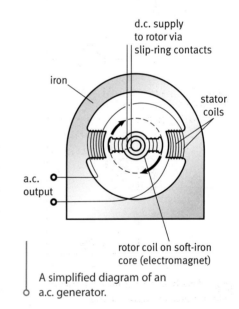

A simplified diagram of an a.c. generator.

Questions

1   Describe the difference between alternating current produced by a power station and the direct current from a battery.

2   Look at the graph. Explain why the voltage across the coil changes with time as the magnet rotates.

Key words

*   ✔ **generator**
*   ✔ **electromagnetic induction**
*   ✔ **alternating current (a.c)**
*   ✔ **direct current (d.c)**

Transformers

An electric current can be generated by moving a magnet into (or out of) a coil of wire. The moving magnet could be replaced by an electromagnet. If a coil is wound around an iron core, it becomes quite a strong magnet when a current flows through it.

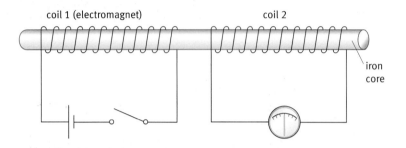

When the current in the electromagnet (coil 1) is switched on, this has the same effect as plunging a bar magnet into coil 2. So a current is generated in coil 2, whilst the current in coil 1 is changing. This arrangement of two coils on the same iron core is called a **transformer**. Changing the current in the primary coil induces a voltage across the secondary coil. If the current in the primary is a.c., it is changing all the time. So an alternating voltage is induced across the secondary coil.

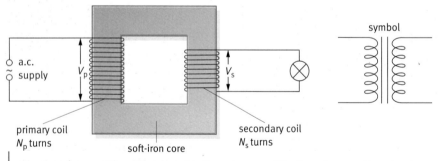

The transformer. When the current in the primary coil is changing, a voltage is induced across the secondary coil. This makes a current flow round the right-hand circuit. Notice that there is no direct electrical connection between the coils of a transformer. The only connection is through the magnetic field.

The output voltage of a transformer depends on the number of turns of wire on the two coils.

$$\frac{\text{voltage across secondary coil } (V_S)}{\text{voltage across primary coil } (V_P)} = \frac{\text{number of turns on secondary coil } (N_S)}{\text{number of turns on primary coil } (N_P)}$$

If there are more turns on the secondary coil, then the induced voltage across this coil is bigger than the applied voltage across the primary coil. However, this is not something for nothing! The current in the secondary will be less, so that the power from the secondary is no greater than the power supplied to the primary (remember: power = IV).

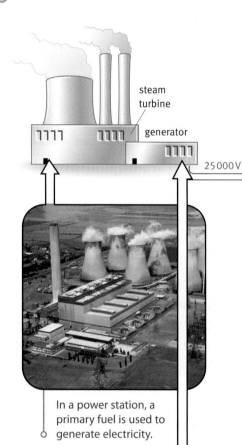

In a power station, a primary fuel is used to generate electricity.

The heart of a power station is a turbine, and as it turns it makes the coil of a generator rotate.

The National Grid

Transformers play an important role in the National Grid system. The National Grid distributes electricity from the power stations to the rest of the country. It does this by means of a long chain of links. These are cables and magnetic fields in transformers.

Transformers are used to step up the output from the power stations and then step down again at the end of the power line. This can only happen because mains electricity supplies an alternating current.

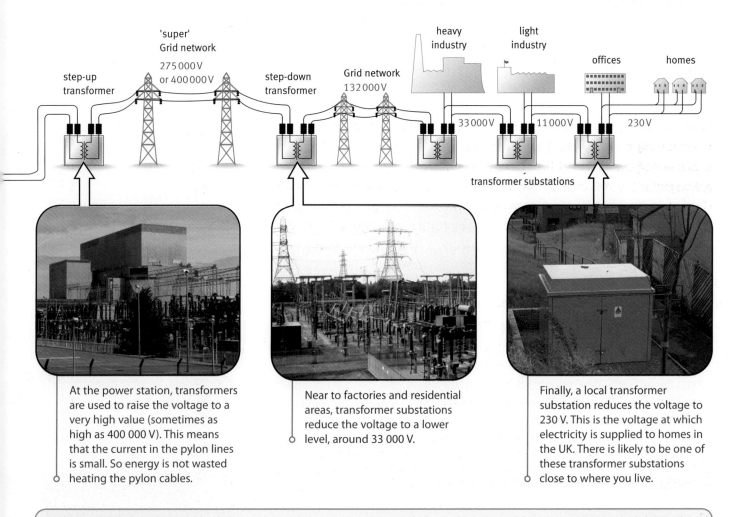

At the power station, transformers are used to raise the voltage to a very high value (sometimes as high as 400 000 V). This means that the current in the pylon lines is small. So energy is not wasted heating the pylon cables.

Near to factories and residential areas, transformer substations reduce the voltage to a lower level, around 33 000 V.

Finally, a local transformer substation reduces the voltage to 230 V. This is the voltage at which electricity is supplied to homes in the UK. There is likely to be one of these transformer substations close to where you live.

Questions

1 A transformer has 100 turns on its primary coil, and 25 turns on its secondary coil. A 12 V a.c. supply is connected to the primary coil. What will be the voltage across the secondary coil?

2 You have a 6 V a.c. supply and want to use it to operate a 12 V bulb. Explain how you could make a simple transformer to do this.

3 In the National Grid, transformers are used to 'step up' the voltage from 25 000 V to 400 000 V. What gets smaller as a result (and stops us getting something for nothing)?

Science Explanations

Electricity is essential to modern-day life. An understanding of electric charge, current, voltage, and resistance in a circuit allows us to use electricity safely and enables power to be generated and distributed.

You should know:

- about electric charge and how positive and negative charges can be separated
- that electric current is a flow of charges already present in the materials of the circuit
- how electric circuits work, and about models that help us understand electric circuits
- that current is not used up as it goes around but it does work on the components it passes through, transferring energy from the battery to other components
- that the voltage of a battery is a measure of the push on the charges
- that the bigger the voltage the bigger the current
- that the components in a circuit resist the flow of charge and how the current depends on the battery voltage and the circuit resistance
- why resistors get hotter when current flows through them and why filament lamps glow
- about components with a variable resistance, including thermistors and light-dependent resistors
- how to measure the voltage between two points using a voltmeter
- that the battery can be thought of as raising charges to a higher level of potential energy and that the charges then lose this energy as they go around the circuit
- that the voltage is also called the potential difference (p.d.)
- about the p.d. across and the current through resistors connected in series and in parallel
- that a circuit with two resistors is sometimes called a potential divider and how this circuit can be useful when used with a variable resistor, thermistor, or LDR
- about the power (energy per second) transferred by an electric circuit
- about the force on a current-carrying wire in a magnetic field and why a motor spins
- about electromagnetic induction, including:
  - a p.d. is induced across the ends of a wire, or coil, in a changing magnetic field
  - if this wire, or coil, is part of a circuit there is an induced electric current in the circuit
  - the magnetic field must be changing, otherwise there is no effect
  - how the effect is increased and how it is used in making an electrical generator
- that electrical generators are used to produce mains electricity
- the difference between a.c. and d.c. electricity
- about the electricity supply system (the National Grid) and why it uses high voltages although the mains voltage to our homes is 230 V.
- about transformers and the effect of changing the number of turns in the coils.

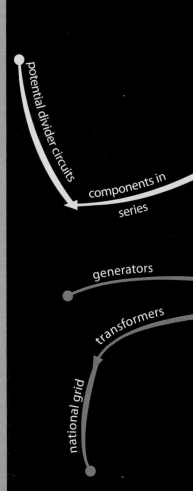

potential divider circuits

components in series

generators

transformers

national grid

ELECTRIC CIRCUITS

variable resistor
Ohm's Law
thermistor
LDR
heating effect
electrical reisistance
filament lamp
size of voltage
using a voltmeter
battery
voltage
size of current
flow of charge
positive and negative charge
charges free to move?
insulators
other conductors
metal conductors
free electrons
electric current
circuit symbols
drawing circuits
plot and interpret graphs
using equations
voltage, current and resistance
power, potential difference,
current and charge
energy, power, time
calculations using data
mean
range
mathematical skills
factors and outcomes
cause
correlation
connecting components
components in parallel
electromagnetic induction
a.c and d.c
mains electricity
magnetic field
electric motors
current carrier
force
energy
power
work done
units
joules
kilowatt hours

Ideas about Science

In addition to developing an understanding of electric circuits, it is important to understand how scientists use data to develop their ideas. Collecting data is often the starting point for a scientific enquiry, but data can never be trusted completely. Data is more reliable if it can be repeated; when making several measurements of the same quantity, the results are likely to vary. This may be because:

- you have measured several individual samples, for example, several samples of a resistance wire
- the quantity you are measuring is varying, for example, the light level in the room is varying as you measure the resistance of an LDR
- there are limitations in the measuring equipment, for example, a poor electrical connection in the circuit.

Usually the best estimate of the true value of a quantity is the mean of several repeated measurements. When there is a spread of values in a set of measurements, the true value is probably in the range between the highest and the lowest values. You should:

- be able to calculate the mean from a set of repeat measurements
- know that a measurement may be an outlier if it is well outside the range of other measurements
- be able to explain whether or not an outlier should be included as part of the data or rejected when calculating the mean.

When comparing sets of data to decide if there is a difference between the two means, it is useful to look at the ranges of the data. You should know:

- if the ranges of two sets of data do not overlap there may be a real difference between the means.

To investigate the relationship between a factor and an outcome, it is important to control all the other factors that might affect the outcome. In a plan for an investigation you should be able to:

- recognise that the control of other factors is a positive feature of an investigation and it is a design flaw if factors are not controlled
- explain why it is necessary to control all the factors that might affect the outcome, other than the factor being investigated, for example, if investigating how the thickness of a wire affects its resistance, use the same material and length for each test.

Factors and outcomes may be linked in different ways, and it is important to distinguish between them. A correlation between a factor and an outcome does not necessarily mean that the factor causes the outcome; both might be caused by some other factor. For example, the more electricity substations there are in an area, the more babies are born in that area. But this is because there are more houses needing an electricity supply where more people live. You should be able to:

- identify a correlation from data, a graph, or a description
- explain why an observed correlation does not necessarily mean that the factor causes the outcome
- explain why individual cases do not provide convincing evidence for or against a correlation.

Review Questions

1 Look at the electric circuit models in this module. Copy and complete the following table.

Model	What corresponds to:		
	the battery?	electric current?	the resistors or lamps?
'peas in a pipe'			
'water in a pipe'			

2 In a simple single-loop electric circuit, the current is the same everywhere. It is not used up. How does each of the models above help to account for this?

3 Imagine a simple electric circuit consisting of a battery and a bulb. For each of the following statements, say if it is true or false (and explain why):

a Before the battery is connected, there are no electric charges in the wire. When the circuit is switched on, electric charges flow out of the battery into the wire.

b Collisions between the moving charges and fixed atoms in the bulb filament make it heat up and light.

c Electric charges are used up in the bulb to make it light.

4 In shops, you can buy batteries labelled 1.5 V, 4.5 V, 6 V, or 9 V. But you cannot buy batteries labelled 1.5 A, 4.5 A, 6 A, or 9 A. Explain why not.

5 You are given four 4 Ω resistors. Draw diagrams to show how you could connect all four together to make a resistance of:

a 16 Ω   **b** 1 Ω

c 10 Ω   **d** 4 Ω

Note that there is more than one possible way to do parts c and d.

6 Peter has a sensor labelled LDR.

a What do the letters LDR stand for?

b What does an LDR detect?

c What does Peter need to measure to work out the resistance?

d Draw a circuit diagram to show how he could measure the quantities in your answer to part b.

7 Copy and complete these sentences:

When a magnet is moved into a coil of wire, a voltage is _____ in the coil. The voltage is produced only when the magnet is _____.
This is used in an a.c. generator, which has an _____ rotating near a fixed coil. To increase the size of the induced voltage, you could use a _____electromagnet, have more _____ on the fixed coil, turn the rotor coil _____, or put a core of _____ inside it.
The current in the external circuit constantly changes direction, so it is called _____ current (__). This is different from the current from a battery, which always goes in one direction and is called current (__).

8 What are the similarities and differences between a motor and a generator?

9 A school laboratory has a set of transformers to demonstrate how power lines work. The transformer has 240 turns on the primary coil and 1200 turns on the secondary coil.

a How will the output voltage be different to the input voltage?

b The input voltage is 2 V. Calculate the output voltage.

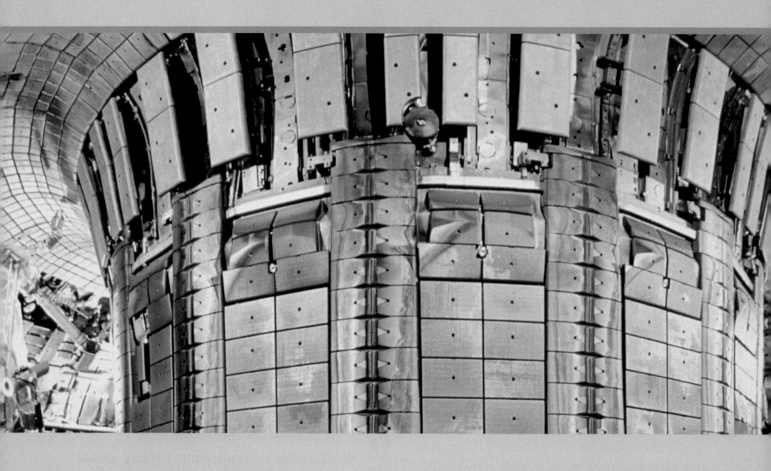

P6 Radioactive materials

Why study radioactive materials?

People make jokes about radioactivity. If you visit a nuclear power station, or if you have hospital treatment with radiation, they may say you will 'glow in the dark'. People may worry about radioactivity when they don't need to. Most of us take electricity for granted. But today's power stations are becoming old and soon will need replacement. Should nuclear power stations be built as replacements? Should we research nuclear fusion as a long-term energy solution?

What you already know

- Some materials are radioactive, and naturally emit gamma rays.
- Gamma rays are ionising radiation.
- Ionising radiation can damage living cells.
- Nuclear power stations produce radioactive waste.
- Contamination by a radioactive material is more dangerous than a short period of irradiation.

Find out about

- radioactive materials and emissions
- radioactive materials being used to treat cancer
- ways of reducing risks from radioactive materials
- nuclear power stations
- nuclear fusion research.

The Science

The discovery of radioactivity changed ideas about matter and atoms. The nuclear model of the atom helped scientists explain many observations – including radioactivity and the colour of stars. Having a model for how the atom behaves enables scientists to make and test predictions, including nuclear fission and nuclear fussion.

Ideas about Science

Radioactive materials are used in many applications. Making decisions about how they are used requires decisions to be made about the balance of risk and benefit, both to the individual but also to the wider community.

Find out about

- radioactive decay
- what makes an atom radioactive
- types of radiation

In this research the scientist is measuring the radioactivity of the fruit from plants grown with radioactive water.

Key words

- radioactive
- ionising radiation
- radioactive decay
- nucleus
- unstable
- alpha particles
- beta particles
- gamma radiation

What do these elements have in common?

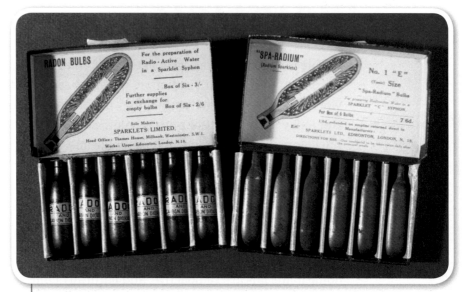

Radon is a radioactive gas. Radium is a radioactive metal. In the early 1900s, these bulbs were used to make drinking water radioactive.

Uranium ore – uranium is used as a fuel in nuclear power stations.

They are all **radioactive**. If you test them with a Geiger counter you will hear it click.

When radioactivity was first discovered, people did not know that the radiation was **ionising** and could damage or kill living cells. They thought that it was natural and healthy. Manufacturers made all kinds of products using radioactive materials. When scientists realised the danger the products were banned. Safety rules for using radioactive materials were introduced.

Radioactive elements can be naturally occurring, or they can be man-made. The man-made elements may be produced because they are useful. For example, the radioactive hydrogen in the water molecules is used to water the plants in the photograph. Or they may be a waste product, like waste from nuclear power stations.

Changes inside the atom

Many elements have more than one type of atom. For example, some carbon atoms are radioactive. In most ways they are identical to other carbon atoms. All can:

- be part of coal, diamond, or graphite
- burn to form carbon dioxide
- be a part of complex molecules.

Radioactive decay

The main difference is that most carbon atoms do not change. They are stable.

Radioactive carbon atoms randomly give out energetic radiation. Each atom does it only once. And what is left afterwards is not carbon, but a different element . The process is called **radioactive decay**. It is not a chemical change; it is a change *inside* the atom.

What makes an atom radioactive?

Atoms have a tiny core called the **nucleus**. In some atoms, the nucleus, is **unstable**. The atom decays to become more stable. It emits energetic radiation and the nucleus changes. This is why the word 'nuclear' appears in *nuclear reactor*, *nuclear medicine*, and *nuclear weapon*.

Three types of radiation are emitted, called alpha, beta, and gamma.

A cut diamond sitting on a lump of coal. Each of these is made of carbon atoms. Some of the atoms will be radioactive.

Radiation	What it is
alpha particles (α)	small, high-speed particle with + charge
beta particles (β)	smaller, higher-speed particle with − charge
gamma radiation (γ)	high-energy electromagnetic radiation

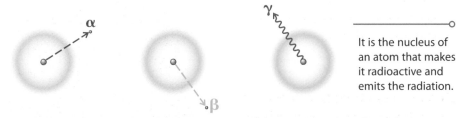

It is the nucleus of an atom that makes it radioactive and emits the radiation.

Radioactive decay happens in the nucleus of an atom. It is not affected by any physical or chemical changes that happen to the atom.

Making gold

When radioactive platinum decays it turns into a new element – gold. A good way to make money?
No. The price of gold is only half the price of platinum.

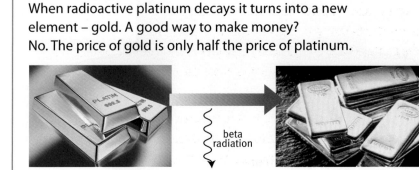

beta radiation

Questions

1 How can you test to see if something is radioactive?

2 Why is ionising radiation dangerous?

3 What part of the atom does the radiation come from?

4 What are the three different types of radiation from radioactive materials?

Find out about

- ✔ **models of the atom**
- ✔ **how alpha particle scattering reveals the existence of the atomic nucleus**

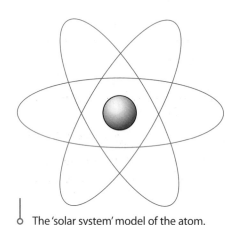

The 'solar system' model of the atom.

How do scientists know about the structure of atoms?

The diagram of an atom as a miniature 'solar system', with the nucleus at the centre and electrons whizzing round like miniature planets, has become very familiar. It is often used just to suggest that something is 'scientific'.

The 'solar system' model of the atom dates back to 1910, and an experiment thought up by Ernest Rutherford. Scientists were beginning to understand radioactivity, and were experimenting with radiation. Rutherford realised that alpha and beta particles were smaller than atoms, and so they might be useful tools for probing the structure of atoms. So he designed a suitable experiment, and it was carried out by his assistants, Hans Geiger and Ernest Marsden.

Here is how to do it:

- Start with a metal foil. Use gold, because it can be rolled out very thin, to a thickness of just a few atoms.
- Direct a source of alpha radiation at the foil. Do this in a vacuum chamber, so that the alpha particles are not absorbed by air.
- Watch for flashes of light as the alpha particles strike the detecting material on the screen at the end of the microscope.
- Work all night, counting the flashes at different angles, to see how much the alpha radiation is deflected.

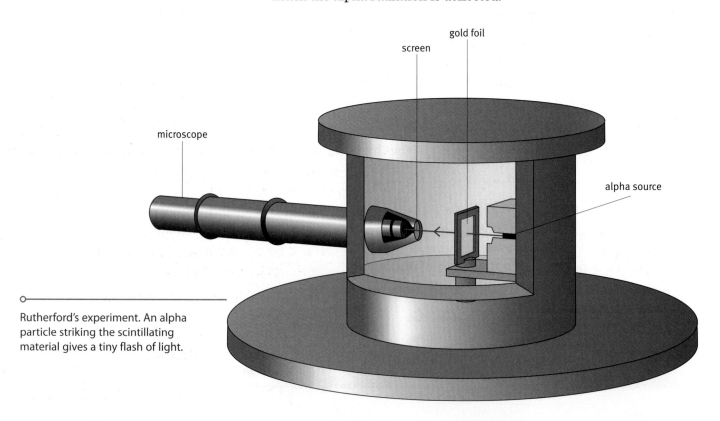

Rutherford's experiment. An alpha particle striking the scintillating material gives a tiny flash of light.

Results and interpretation

This is what Geiger and Marsden observed:

- Most of the alpha particles passed straight through the gold foil, deflected by no more than a few degrees.
- A small fraction of the alpha particles were actually reflected back towards the direction from which they had come.

And here is what Rutherford said:
'It was as if, on firing a bullet at a sheet of tissue paper, the bullet were to bounce back at you!'

In fact, less than 1 alpha particle in 8000 was back-scattered (deflected through an angle greater than 90°), but it still needed an explanation.

Rutherford realised that there must be something with positive charge that was repelling the alpha particles (which also have positive charge). And it must also have a lot of mass, or the alpha particles would just push it out of the way.

This 'something' is the nucleus of a gold atom. It contains all of the positive charge within the atom, and most of the mass. Rutherford's nuclear model is a good example of a scientist using creative thinking to develop an explanation of the data.

His analysis of his data showed that the nucleus was very tiny, because most alpha particles flew straight past without being affected by it. The diameter of the nucleus of an atom is roughly a hundred-thousandth of the diameter of the atom.

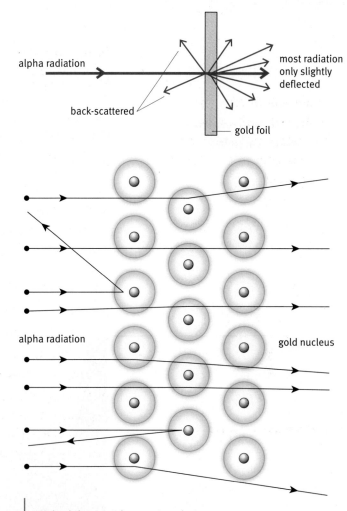

Only alpha particles passing close to a nucleus are significantly deflected.

Key word

✓ **alpha particle scattering**

Questions

1 What charge do the following have:
   a the atomic nucleus?
   b alpha radiation?
   c electrons?

2 Put these in order, from least mass to greatest: gold atom, alpha particle, gold nucleus, electron.

3 Describe and explain what happened to alpha particles that were directed:
   a straight towards a gold nucleus
   b slightly to one side of a gold nucleus
   c midway between two nuclei.

4 Suggest how the results would be different if the nucleus:
   a was positive but very large
   b was negative and very small.

Find out about

- ✔ isotopes
- ✔ protons and neutrons
- ✔ α and β particles

carbon-12

carbon-11

The nucleus of carbon-12 has 6 protons and 6 neutrons. Carbon-11 has 11 particles in its nucleus: 6 protons and 5 neutrons.

Compared to the whole atom, the tiny nucleus is like a pinhead in a stadium.

Question

1 Look at these isotopes: carbon-11, boron-11, carbon-12, nitrogen-12.

  a Which two are isotopes of the same element?

  b Which ones have the same number of particles in the nucleus?

  c Do any of them have identical nuclei?

  d A nucleus of carbon-14 has **i** how many protons?

  ii how many neutrons?

Atoms are small – about a ten millionth of a millimetre across. Their outer layer is made of electrons. Most of their mass is concentrated in a tiny core, called a nucleus.

Isotopes

The tiny nucleus contains two types of particle: **protons** and **neutrons**. All atoms of any element have the same number of protons. For example, carbon atoms always have six protons. But they can have different numbers of neutrons and still be carbon. The word **isotope** is used to describe different atoms of the same element. Carbon-11 and carbon-12 are different isotopes of carbon.

Carbon-11 will give out its radiation whether it is in diamond, coal, or graphite. You can burn it or vaporise it and it will still be radioactive.

Describing a nucleus

Scientists use a formula for describing an isotope.

In front of the chemical symbol are written the **proton number** and the total number of particles (protons + neutrons) in the nucleus.

$^{11}_{6}\text{C}$

This is the number of particles in the nucleus

This is the number of protons

The number of neutrons can be found by taking the number of protons away from the total number of particles in the nucleus.

Neutrons = 11 − 6 = 5

$^{12}_{6}\text{C}$

protons + neutrons

protons

neutrons = 12 − 6 = 6

Radioactive changes

Some nuclei that are **unstable** can become more stable by emitting an alpha particle. An alpha particle is made of two protons and two neutrons – it is the same as a helium nucleus.

Other nuclei can become more stable when a neutron decays to form a proton. It does this by emitting a beta particle. A beta particle is the same as an electron, but it has come from a neutron in the nucleus. It is not one of the atom's orbital electrons.

Counting the particles

Scientists use nuclear equations to work out what happens during radioactive decay.

Alpha decay

When plutonium-240 decays uranium-236 is formed.

$$^{240}_{94}\text{Pu} \longrightarrow {}^{336}_{92}\text{U} + {}^{4}_{2}\alpha$$

Beta decay

When carbon-14 decays nitrogen-14 is formed. Nitrogen always has 7 protons in the nucleus.

$$^{14}_{6}\text{C} \longrightarrow {}^{14}_{7}\text{N} + {}^{0}_{-1}\beta$$

The emission of either an alpha or a beta particle from an unstable nucleus produces an atom of a different element, called a 'daughter product' or 'decay product'. The daughter product may itself be unstable. There may be a series of changes, but eventually a stable end-element is formed.

α particle $= {}^{4}_{2}\text{He}$

An alpha particle has two protons and two neutrons.

β particle

A neutron decays to a proton and an electron. The electron is a beta particle.

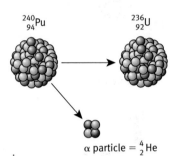
$^{240}_{94}\text{Pu}$ $^{236}_{92}\text{U}$

α particle $= {}^{4}_{2}\text{He}$

During alpha decay the nucleus loses two protons and two neutrons.

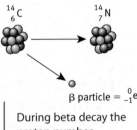
$^{14}_{6}\text{C}$ $^{14}_{7}\text{N}$

β particle $= {}^{0}_{-1}\text{e}$

During beta decay the proton number increases by one.

Isotope	Number of protons	Radioactive?
Hydrogen-1	1	
Hydrogen-2	1	
Hydrogen-3	1	✓
Helium-3	2	✓
Helium-4	2	
Uranium-235	92	✓
Thorium-231	90	✓

Questions

2   Copy the table of isotopes and add three extra columns for:
   • the total number of particles in the nucleus
   • the number of neutrons in the isotope
   • the chemical symbol for the isotope.
   Add the information for each of these columns to complete the table.

3   Hydrogen-3 decays to helium-3 by beta decay. Write a nuclear equation for the decay.

4   Uranium-235 decays to thorium-231 by alpha decay. Write a nuclear equation for the decay.

5   Nitrogen-16 decays by beta decay. Use your answer to question three to write a nuclear equation for the decay.

Key words

✓ **isotope**
✓ **proton**
✓ **neutron**
✓ **unstable**
✓ **proton number**

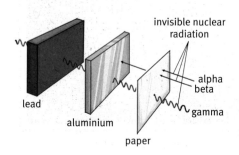

Radioactive isotopes have many uses, but they are quite rare in nature – because most of them have decayed – so radioactive isotopes are made in nuclear reactors, or in accelerators, and are prepared for use in laboratories and hospitals around the country.

Alpha, beta, or gamma?

A beam of alpha particles may be the best choice for the job, or it might be beta particles, or gamma rays. To decide, scientists consider the properties described below.

Alpha radiation

Alpha particles are much heavier than beta particles, and they quickly collide with air molecules and slow down. They gain electrons and become atoms of helium. This means that they are the least penetrating, but also the most strongly ionising radiation. They are stopped most easily.

Beta radiation

Beta particles are fast moving electrons. They are much smaller than alpha particles so less likely to collide with other particles. This means they travel further in air and other materials and are less ionising. When they have slowed down they are just like any other electrons.

Gamma radiation

Sometimes, after a nucleus emits a beta particle, the protons and neutrons remaining in the new nucleus rearrange themselves to a lower energy state. When this happens the nucleus emits a photon of electromagnetic radiation called a gamma ray. This does not cause a change of element. The photons have more energy than most X-ray photons and rarely collide with particles, so the radiation is very penetrating. It has only a very weak ionising effect.

Question

1 Which type of radiation:
   a is the most penetrating?
   b is the most ionising?
   c has the longest range in air?

Radiation	Range in air	Stopped by	Ionisation	Charge
alpha	a few cm	paper / dead skin cells	strong	+
beta	10–15 cm	thin aluminium	weak	–
gamma	many metres	very thick lead or several metres of concrete	very weak	no charge

Properties of ionising radiation.

Sterilisation

Ionising radiations can kill bacteria. Gamma radiation is used for **sterilising** surgical instruments and some hygiene products such as tampons. The products are first sealed from the air and then exposed to the radiation. This passes through the sealed packet and kills the bacteria inside.

Food can be treated in the same way. Irradiating food kills bacteria and prevents spoilage. As of 2010, **irradiation** is permitted in the UK only for herbs and spices. But the label must show that they have been treated with ionising radiation. This is a useful alternative to heating or drying, because it does not affect the taste.

Gamma rays kill the bacteria on and inside these test tubes.

Questions

2 a Why seal the packets of surgical instruments before sterilising them?

b Does the gamma radiation make them radioactive? Explain your answer.

3 Smoke detectors used in homes contain a source that emits alpha particles.

a Explain why these are not dangerous in normal use.

b What might make them dangerous?

4 A radioactive source is tested by placing sheets of material between it and a Geiger counter 2 cm away.

Sheet added	Count rate (counts per second)
None	6.8
Paper	4.9
3-mm-thick aluminium	4.7
3-cm-thick lead	0.5

Which type, or types, of radiation does the source emit? Explain your reasoning.

Food irradiation is it safe?

The logo shows that the herbs and spices have been **irradiated** with gamma radiation from Cobalt–60. Gamma rays pass through the glass and kill any bacteria in the jar. Cobalt–60 does not pass in to the jar– there is *no* **contamination**.

Uses of ionising radiation are linked to their properties.

Key words

✓ irradiation
✓ sterilisation

Find out about

- ✔ **background radiation**
- ✔ **a radioactive gas called radon**
- ✔ **radiation dose and risk**

Radiation sources

If you switch on a Geiger counter, you will hear it click. It is picking up **background radiation**, which is all around you. Most background radiation comes from natural sources.

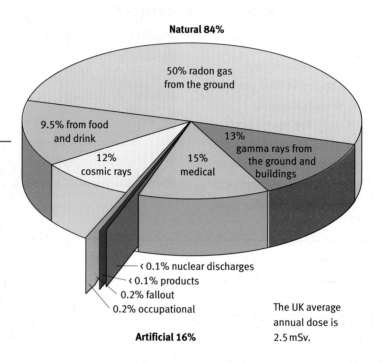

Natural 84%

50% radon gas from the ground

9.5% from food and drink

12% cosmic rays

15% medical

13% gamma rays from the ground and buildings

‹ 0.1% nuclear discharges
‹ 0.1% products
0.2% fallout
0.2% occupational

The UK average annual dose is 2.5 mSv.

Artificial 16%

How different sources contribute to the average **radiation dose** in the UK. Source HSE.

Radiation dose

Radiation dose measures the possible harm the radiation could do to the body. It is measured in millisieverts (mSv). The UK average annual dose is 2.5 mSv. For comparison, with a dose of 1000 mSv (400 times larger) three out of a hundred people, on average, develop a cancer.

Ionising radiation from outer space is called cosmic radiation. Flying to Australia gives you a dose of 0.1 mSv, from cosmic rays. That's not much if you go on holiday, but it soon adds up for flight crews making repeated journeys.

What affects radiation dose?

The dose measures the potential harm done by the radiation. It depends on:

- the amount of radiation – the number of alpha particles, beta particles, or gamma photons reaching the body
- the type of radiation. Alpha is the most ionising of the three radiations. So it can cause the most damage to a cell. The same amount of alpha radiation gives a bigger dose than beta or gamma radiation. But because alpha radiation has such a short range in air, it is only a hazard if the source of radiation gets inside the body.

Questions

1 a In what units is radiation dose measured?
   b Make a reasoned estimate for the annual dose of a long-haul airline pilot.

2 On what two factors does radiation dose depend?

Radiation	Dose factor
alpha	20
beta	1
gamma	1

The damage to the body will also depend on the type of tissue affected. Lung tissue, for example, is easily damaged. Radon gas is dangerous because it emits alpha particles. If it is breathed into the lungs then the alpha radiation will be absorbed in the lung tissue.

Why is ionising radiation dangerous?

Ionising radiation has the energy to break molecules in the cells in the body into ions. These ions can then take part in chemical reactions that might damage the body. If the ionising radiation affects DNA molecules, this may cause the cell to be killed or to behave abnormally. Cells that behave abnormally can cause cancer.

Is there a safe dose?

There is no such thing as a safe dose. Just one radon atom might cause a cancer, just as a person might get knocked down by a bus the first time they cross a road. The chance of it happening is low, but it still exists. The lower the dose, the lower the risk. But the risk is never zero.

Irradiation and contamination

Exposure to a radiation source outside your body is called **irradiation**. Alpha irradiation presents a very low risk because alpha particles:

*   only travel a few centimetres in air
*   are easily absorbed.

Your clothes will stop alpha particles. So will the outer layer of dead cells on your skin.

Irradiation by beta particles is more risky as they penetrate a few centimetres into the body. Most gamma rays pass straight through the body. But they have high energy, so if they are absorbed they are dangerous.

If a radiation source enters your body, or gets on skin or clothes, it is called **contamination**. You become contaminated. If you swallow or breathe in any radioactive material, your vital organs will be exposed to continuous radiation for a long period of time. Sources that emit alpha particles are the most dangerous because alpha particles are the most ionising. Contamination by gamma sources is the least dangerous as most gamma rays will pass straight out of the body.

> **Dose summary**
>
> Radiation dose is affected by:
> *   amount of radiation
> *   type of radiation.

It is difficult to be sure about the harm that low doses of radiation can cause. In the 1970s, Alice Stewart studied the health of people working in the American nuclear industry. Her early results suggested that radiation is more harmful to children and to elderly people. She was attacked for her ideas, and the employers prevented any further access to medical records.

Key words

*   ✔ **background radiation**
*   ✔ **radiation dose**
*   ✔ **irradiation**
*   ✔ **contamination**

Radiation protection

Scientists study radiation hazards and give advice to protect against them. They also keep a close eye on the many people who regularly work with radioactive materials – in hospitals, industry, and nuclear installations. These people are called 'radiation workers'.

Employers must ensure that radiation workers receive a radiation dose 'as low as reasonably achievable'.

This principle applies when better equipment or procedures can reduce the risks of an activity. Any extra cost this involves must be balanced against the amount by which the risk is reduced.

To reduce their dose, medical staff take a number of precautions. They:

* use protective clothing and screens
* wear gloves and aprons
* wear special devices to monitor their dose.

The principle applies equally to hospital patients who receive radiation treatment. If doctors find that one hospital uses smaller doses but is just as effective as any other, then all hospitals are encouraged to copy them.

People working with radiation wear a personal radiation monitor that keeps track of their exposure to ionising radiation.

Staff handle radioactive sources with care to reduce the risk of contact with the source.

Questions

3 Explain the difference between irradiation and contamination.

4 a How big a dose of radiation do you get by catching a flight to Australia?
  b Where do cosmic rays come from?
  c Is this irradiation or contamination?

5 Imagine that alpha radiation damages a cell on the outside of your body. Why is this less risky than internal damage? Give two reasons.

6 In our food there is some potassium-40 (K-40). This is a naturally occurring radioactive isotope that sometimes decays to calcium (Ca) by emitting a beta particle. Write a nuclear equation for this decay. (Potassium has 19 protons in the nucleus.)

7 Explain how each of the precautions that medical staff take helps to keep their radiation dose as low as possible.

Radon gas

Over 400 years ago, a doctor called Georgius Agricola wrote about the high death rate amongst German silver miners. He thought they were being killed by dust, and called their disease 'consumption'.

We now know that the mines contained a high concentration of radon gas. Radon gas is harmful because it is **radioactive**. It produces **ionising radiation** that can damage cells. The silver miners were dying of lung cancer.

Radon and lung cancer

Radon seeps into houses in some areas of the UK, as described in the leaflet on the following page. When this was realised, scientists were concerned about whether low doses over a long time could cause lung cancer. Since then they have conducted lots of studies to see if radon in homes increases the risk of lung cancer.

Scientists measure radon levels in the homes of people with lung cancer and compare them with levels in homes of people who have not got lung cancer. One study compared 413 women with lung cancer with 614 without lung cancer and showed a link between radon exposure and lung cancer. The women had lived in the same homes for 20 years. Some studies have not shown a link. This may be because they had a smaller sample size, or because it is difficult to measure radon exposure over time, especially if people move. One study got round this by analysing glass to measure radon exposure. The scientists chose glass from a mirror or picture frame that was at least 15 years old and had been with the person all the time, even if they moved house. This study did show a correlation between radon exposure and lung cancer.

Find out about

✓ radon gas
✓ radiation dose and risk

Silver mines were contaminated with radon gas. The miners breathed it in and suffered.

The pipe runs beneath the floor of the house and a small fan sucks the radon from the building.

Questions

1 a What was the correlation that Georgius Agricola observed?

  b What do we now know was **i** the factor and **ii** the outcome in what happened to the silver miners?

  c Explain whether the factor *caused* the outcome by discussing the mechanism involved.

2 If you were setting up a study on radon in houses and the risk of lung cancer, explain how you would choose the samples.

A hazard at home

Radon gas builds up in enclosed spaces. In some parts of the UK, it seeps into houses.

LIVING WITH RADON

GOVERNMENT INFORMATION LEAFLET

There is radon all around you. It is radioactive and can be hazardous – especially in high doses.

Radon gives out a type of ionising radiation called **alpha radiation**. Like all ionising radiations, alpha radiation can damage cells and might start a cancerous growth.

Radon is a gas that can build up in enclosed spaces. Some homes are more likely to be contaminated with radon.

What about my home?

You and your family are at risk if you inhale radon-contaminated air. The map shows the areas where there is most contamination.

If you live in one of these areas, get your house tested for radon gas.

What if the test shows radon?

Radon comes from the rocks underneath some buildings. It seeps into unprotected houses through the floorboards. If your house is contaminated, get it protected. An approved builder will put in:
- a concrete seal to keep the radon under your floorboards and
- a pump to remove it safely.

The risk is real: put in a seal.

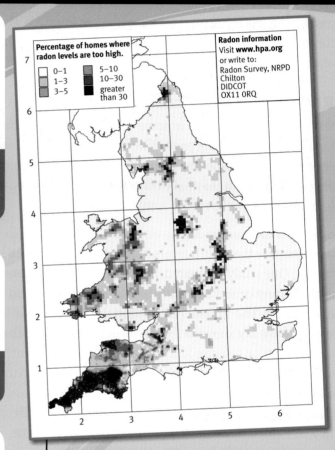

Radon-affected areas in England and Wales. Based on measurements made in over 400 000 homes.

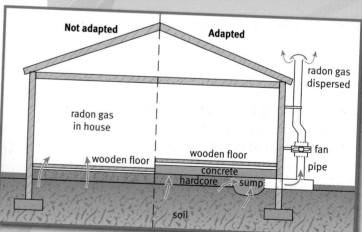

Radon gas can build up inside your home. Sealing the floor and pumping out the gas is an effective cure.

Radon and risk

The risk to miners was high because radon can build up in enclosed spaces, such as mines. In the atmosphere, the radon spreads out. It's a different story in enclosed spaces like mines. The rocks keep producing the gas and it cannot escape. So the radon concentration can be 30000 times higher than in the atmosphere.

There would be a lower concentration in a house, especially if the windows are open or it is draughty. Now that we are careful to draught proof our homes, it is more important to protect them against radon.

On average, radon makes up half the UK annual radiation dose. About 1100 people die each year from its effects, or about 1 in every 50 000 people. But radon is only one of the hazards that people face every day. There are risks associated with driving to school, sunbathing, swimming, catching a plane, and even eating.

The table shows how the risk of cancer from radon compares with some other common risks.

Many risky activities have a benefit. You need to decide whether the risk is worth taking.

Cause of death	Average number of deaths per year
cancer caused by radon	1100
cancer caused by asbestos	4000
skin cancer caused by ultraviolet radiation	1400
road deaths	2500
cancer caused by smoking	35 000
CJD	98
house fire	360
all causes	510 000

Estimated deaths per year in the UK population of 60 million (2008).

Questions

3 In the radon studies, how did analysing the glass of a mirror or picture frame improve the measurement of radon exposure?

4 There is a risk from radon gas building up in houses. Which of these are good ways to reduce the risk?
- stop breathing
- wear a special gas mask
- move house
- adapt the house

5 Choose three causes of death from the table on the left. Write down a way of reducing the risk from each chosen cause on the left.

6 Write a letter to a friend living in a high-radon area to persuade them to get their house checked for radon.

Find out about

✓ the half-life of radioactive materials

Radioactive decay is random. You can never tell which nucleus will decay next. Scientists can't predict whether a particular nucleus will decay today or in a thousand years, time. But in a sample of radioactive material there are billions of atoms, so they can see a pattern in the decay.

The pattern of radioactive decay

The amount of radiation from a radioactive material is called its **activity**. This decreases with time.

• At first there are a lot of radioactive atoms.
• Each atom gives out radiation as it decays to become more stable.
• The activity of the material falls because fewer and fewer radioactive atoms remain.

When a sample of the medical tracer technetium-99m is injected into a patient at 9.00 am then on average, by 3.00 pm half of the nuclei have decayed. At 9.00 pm half of those nuclei remaining will have decayed, leaving only one quarter of the original sample. We say the **half-life** of technetium-99m is six hours. The half-life is the time it takes for the activity to drop by half. The table shows that after 24 hours (four half-lives), only one sixteenth of the original sample remains.

Time	Hours since injection	Number of half-lives	Fraction of original sample remaining
9.00 am	0	0	1
3.00 pm	6	1	$\frac{1}{2}$
9.00 pm	12	2	$\frac{1}{4}$
3.00 am	18	3	$\frac{1}{8}$
9.00 am	24	4	$\frac{1}{16}$

The radioactive decay of Technetium-99m.

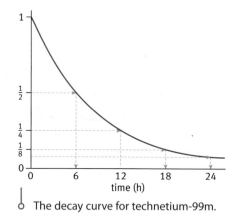
The decay curve for technetium-99m.

Key words

✓ activity
✓ half-life

The graph shows how the activity of the sample decreases. After ten half-lives only about one thousandth of the original sample remains. For technetium-99m this is only two and a half days. The six-hour half-life makes it an ideal isotope for a tracer. It lasts long enough for doctors to get some scans of the decay, but it has almost all gone in a few days.

Different half-lives

All radioactive materials show the same pattern but they can have different half-lives. This graph shows the pattern of radioactive decay for radon. Notice that the amount of radiation halves every minute.

The **half-life** of radon-220 is one minute.

There is no way of slowing down or speeding up the rate at which radioactive materials decay. Some decay slowly over thousands of millions of years. Others decay in milliseconds – less than the blink of an eye.

The shorter the half-life, the greater the activity for the same amount of material. Of the four radioactive isotopes listed in the table, neon-17 is the most active.

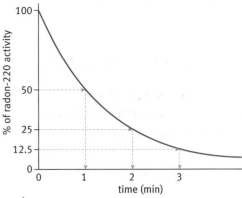

The decay curve for radon-220.

Isotope	Half-life
Iridium-192	74 days
Strontium-81	22 minutes
Uranium-235	710 million years
Neon-17	0.1 seconds

Half-lives can be short or long.

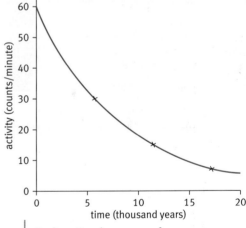

Radioactive decay curve for carbon-14.

Questions

1 Carbon-14 has a half-life of 5700 years. What fraction of its original activity will a sample have after 11 400 years?

2 Iodine-123 is used to investigate problems with the thyroid gland, which absorbs iodine. It is a gamma emitter.
   a Explain why the element iodine is chosen.
   b Explain why it is useful that iodine-123 gives out gamma radiation.
   c Iodine-123 has a half-life of 13 hours. Why would it be a problem if the half-life was:
      i a lot shorter?
      ii a lot longer?

3 How long does it take for a sample of strontium-81 to decay to one eighth of its original size?

4 An underfloor fan is switched on to prevent any more radon-220 entering a house. The owner wants to know how long will it take for the radioactivity from the radon-220 in the house to fall to less than one thousandth of the original value.
   a How many half-lives will this take?
   b How long will this take?

5 The activity of a neon-17 source is 1120 decays per second. What will it be after 0.5 seconds?

6 The activity of an iridium-192 sample is 9600 decays per second. How long will it take to fall to 2400 decays per second?

Find out about

- ✔ **different uses of radiation**
- ✔ **types of radiation**
- ✔ **benefits and risks of using radioactive materials**
- ✔ **limiting radiation dose**

Radioactive materials can cause cancer. But they can also be used to diagnose and cure many health problems.

Medical imaging

Jo has been feeling unusually tired for some time. Her doctors decide to investigate whether an infection may have damaged her kidneys when she was younger.

They plan to give her an injection of DMSA. This is a chemical that is taken up by normal kidney cells.

The DMSA has been labelled as radioactive. This means its molecules contain an atom of technetium-99m (Tc-99m), which has a half-life of six hours. The kidneys cannot tell the difference between normal DMSA and labelled DMSA. They absorb both types.

The Tc-99m gives out its gamma radiation from within the kidneys. Gamma radiation is very penetrating, so nearly all of it escapes from Jo's body and is picked up by a gamma camera. The gamma camera **traces** where the technetium goes in Jo's body. Parts of the kidney, which are working normally, will appear to glow. Any dark or blank areas show where the kidney isn't working properly.

Jo's scan shows that she has only a small area of damage. The doctors will take no further action.

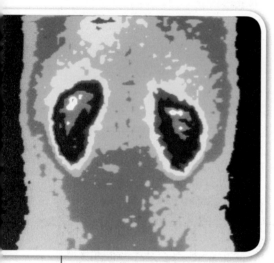

This gamma scan shows correctly functioning kidneys – the top two white areas.

Glowing in the dark

Jo was temporarily contaminated by the radioactive Tc-99m. For the next few hours, until her body got rid of the technetium, she was told to:
- flush the toilet a few times after using it
- wash her hands thoroughly
- avoid close physical contact with friends and family.

Is it worth it?

There was a small chance that some gamma radiation would damage Jo's healthy cells. Before the treatment, her mum had to sign a consent form, and the doctors checked that Jo was not pregnant.

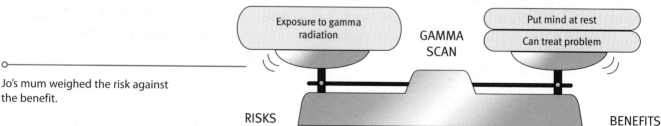

Jo's mum weighed the risk against the benefit.

RISKS — Exposure to gamma radiation — GAMMA SCAN — Put mind at rest / Can treat problem — BENEFITS

Jo's mum said 'We felt the risk was very small. It was worth it to find out what was wrong. Even with ordinary medicines, there can be risks. You have to weigh these things up. Nothing is completely safe.'

Treatment for thyroid cancer

Alf has thyroid cancer. First he will have surgery, to remove the tumour. Then he must have **radiotherapy**, to kill any cancer cells that may remain.

A hospital leaflet describes what will happen.

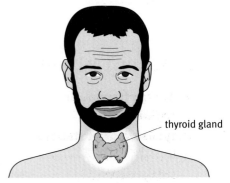

The thyroid gland is located in the front of the neck, below the voice box.

Radioiodine treatment

You will have to come in to hospital for a few days. You will stay in a single room.

You will be given a capsule to swallow, which contains iodine-131. This form of iodine is radioactive. You cannot eat or drink anything else for a couple of hours.

- The radioiodine is absorbed in your body.
- Radioiodine naturally collects in your thyroid, because this gland uses iodine to make its hormone.
- The radioiodine gives out beta radiation, which is absorbed in the thyroid.
- Any remaining cancer cells should be killed by the radiation.

You will have to stay in your room and take some precautions for the safety of visitors and staff. You will remain in hospital for a few days, until the amount of radioactivity in your body has fallen sufficiently.

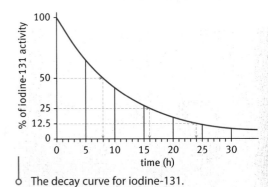

The decay curve for iodine-131.

Key words
- ✓ **radiotherapy**
- ✓ **traces**

Iodine-131 decays to an isotope of xenon.

Questions

1 Look at the precautions that Jo has to take after the scan. Write a few sentences explaining to Jo why she has to do each of them.

2 It would be safe to stand next to Jo but not to kiss her. Use the words 'irradiation' and 'contamination' to explain why.

3 What are the risks and the benefits to Jo of having the treatment?

4 Read the information leaflet about radioiodine. Describe how the risk to Alf's family and other patients is kept as low as possible.

5 Radioiodine has a half-life of eight days. Explain why a half-life of eight days is more suitable than:
   a eight minutes
   b eight years.

6 Alf has a check up after 40 days.
   a How many half-lives of iodine 131 is 40 days?
   b What fraction of the radiation remains after 40 days?

Nuclear fission

Radioactive atoms have an unstable nucleus. Some nuclei can be made so unstable that they split in two. This process is called **nuclear fission**.

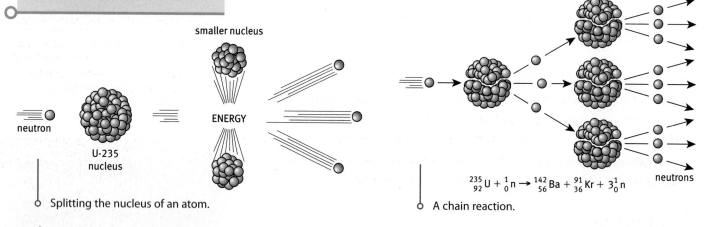

smaller nucleus

ENERGY

neutron

U-235 nucleus

Splitting the nucleus of an atom.

$$^{235}_{92}U + ^{1}_{0}n \rightarrow ^{142}_{56}Ba + ^{91}_{36}Kr + 3^{1}_{0}n$$

neutrons

A chain reaction.

For example, the nucleus of a uranium-235 atom breaks apart when it absorbs a neutron.

When this happens a small amount of the mass of the nucleus is converted to a huge amount of energy. The energy released can be calculated from **Einstein's equation**:

$$E = mc^2$$

where E = energy, m = change in mass and c = speed of light.

The energy is transferred to the fission products, so they have a lot of kinetic energy. Each fission reaction produces roughly a million times more energy than when a molecule changes during a chemical reaction.

The fission of one atom can set off several more, because each fission reaction releases a few neutrons. If there are enough U-235 atoms close together, there will be a **chain reaction**, involving more and more atoms.

The devastating power of a nuclear weapon.

Nuclear weapons

In World War 2 there was a race to 'split the atom' and harness the energy in the form of a bomb.
On 16 July 1945, in the deserts of New Mexico, a group of scientists waited tensely as they tested 'the gadget'. Some thought it would be a flop. Others worried that it might destroy the atmosphere.
At 5.29 a.m., it was detonated and filled the skies with light. The bomb vaporized the metal tower supporting it. All desert sand within a distance of 700 m was turned into glass.
Some of the scientists were worried about the power of the bomb and wanted the project stopped. A few weeks later, the Americans dropped two nuclear bombs on Japan, at Hiroshima and Nagasaki.

Controlling the chain

At the heart of a nuclear power station is a reactor. It is designed to release the energy of uranium at a slow and steady rate, by controlling a chain reaction.

- The fission takes place in **fuel rods** that contain uranium-235. This makes them extremely hot.
- **Control rods**, which contain the element boron, absorb neutrons. Moving control rods in or out of the reactor decreases or increases the reaction rate.

Generating electricity

A fluid, called a **coolant**, is pumped through the reactor. The hot fuel rods heat the coolant to around 500 °C. It then flows through a heat exchanger in the boiler, turning water into steam. The steam is used in the same way as in a coal-fired power station. One reason for building nuclear power stations is to reduce the need for fossil fuels.

Plutonium

The element plutonium is produced in nuclear reactors. It can also undergo nuclear fission so it can be used for nuclear weapons or fuel. Countries sometimes build nuclear reactors to obtain plutonium for weapons. Nuclear weapons inspectors try to ensure that nuclear power stations are very secure, account for all their waste, and are not operated in unstable countries.

The Nuclear Installations Inspectorate monitors the design and operation of nuclear reactors. Reactor cores are sealed and shielded. Very little radiation gets out.

Once fuel rods are in service, decay products build up in them. They become more radioactive.

Questions

1 Why do nuclear reactors use coolants and not circulate water to make steam directly?

2 a Complete the nuclear equation for the absorption of a neutron by Uranium-235:

$$^{235}_{92}\text{U} + ^{0}_{1}\text{n} \longrightarrow ^{:::}_{:}\text{U}$$

b Describe what happens to this unstable nucleus of uranium and how this can lead to a chain reaction.

c If the chain reaction runs out of control we have an atomic bomb. Explain how it is controlled in a nuclear reactor.

3 Suggest how gamma radiation from a nuclear reactor is contained, so that living things are not irradiated.

4 Write down two risks and two benefits of living in a country with nuclear power stations.

Key words

- ✓ fuel rod
- ✓ control rod
- ✓ coolant

Find out about

- the UK's nuclear legacy
- the half-life of radioactive materials
- possible methods of disposal

High-level radioactive waste is hot, so it is stored underwater.

The control room at a nuclear waste storage plant enables people to monitor the waste continuously.

A legacy of nuclear waste

The Nuclear Decommissioning Agency (NDA) is responsible for cleaning up hazardous nuclear waste at 36 sites around the UK. These include power stations, research sites, Ministry of Defence sites and healthcare sites. Most of the radioactive waste comes from power stations. The rest comes from medical uses, industry, and scientific research. In addition to 'everyday' waste, when power stations are too old to be used anymore they must be safely dismantled. The waste radioactive materials are separated out and taken away to be stored. Nuclear waste is a cocktail made of different isotopes. They call it the UK's 'nuclear legacy'.

A long-term hazard

Radioactive waste has very little effect on the UK's average background radiation. But it is still hazardous. This is because of contamination. Imagine that some waste leaks into the water supply. This could be taken up by food, which you eat. The radioactive material is now in your stomach, where it can irradiate your internal organs.

Some radioactive materials have half-lives of thousands of years. The NDA must dispose of nuclear waste in ways that are safe and secure for many, many generations.

Types of waste

The nuclear industry deals with three types of nuclear waste.

- **High-level waste** (HLW). This is 'spent' fuel rods. HLW gets hot because it is so radioactive. It has to be stored carefully but it doesn't last long. And there isn't very much of it. All the UK's HLW is kept in a pool of water at Sellafield.

- **Intermediate-level waste** (ILW). This is less radioactive than HLW. But the amount of ILW is increasing, as HLW decays to become ILW.

- **Low-level waste** (LLW). Protective clothing and medical equipment can be slightly radioactive. It is packed in drums and dumped in a landfill site that has been lined to prevent leaks.

Type of waste	Volume (m³)	Radioactivity
LLW	196 000	weak
ILW	92 500	strong
HLW	1730	extremely strong

The amount of nuclear waste in store (2007). The problem of what to do with it remains unsolved.

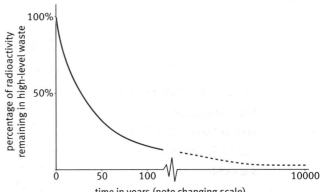

high-level waste decays quickly at first. When its activity falls, it becomes intermediate-level waste. ILW stays radioactive for thousands of years.

Sellafield

A government-owned company runs the biggest nuclear site in the UK – Sellafield, in Cumbria. Thousands of workers – professional, skilled, and unskilled – contribute to its important work. Sellafield reprocesses nuclear waste produced in the UK and abroad. It also prepares and stores nuclear waste for permanent disposal.

Risk management is a major concern at Sellafield. They must plan in advance how to maintain production and safety in the event of any possible problem.

ILW presents the biggest technical challenge, because it is very long-lived. Currently it is chopped up, mixed with concrete, and stored in thousands of large stainless-steel containers. This is secure but not permanent. The long-term solution has to be secure and permanent.

The work of the NDA

Managing waste is very expensive. In 2010 the NDA spent £28 billion on work at 19 sites including Sellafield. One priority is to complete the reprocessing of the HLW at Sellafield by 2016. In 2007 an inventory was carried out of all the radioactive material stored at the 36 sites. There have been a number of public consultations about what to do with the waste. At the time of writing these are still going on. The preferred plan at the moment is to store it until a safe geological site can be found and then to bury it.

When will it be 'safe'?

We are exposed to some radiation all the time (background radiation). When the nuclear waste only emits very low levels of radiation, similar to the background radiation, it poses little risk. The longer the half-life of the radioactive material, the longer it will take to become 'safe'.

Key words
- high-level waste
- intermediate-level waste
- low-level waste

Questions

1 Disposing of ILW needs to be both secure and permanent. Explain why both criteria are important.

2 The NDA wants to know the public's views on waste storage. Write a letter explaining what you think should be done.

3 What are the advantages and disadvantages of keeping all the waste together above ground rather than burying it in a deep shaft and sealing it?

4 A small amount of nuclear fuel produces a lot of energy, so in the 1950s scientists thought this would be a cheap way of generating electricity. Explain why the real cost is much greater than realised at the time.

5 HLW contains plutonium-239, which has a half-life of 24 100 years. Explain some of the difficulties of keeping it safe for 10 half-lives.

Find out about

- ✔ **nuclear fusion**
- ✔ **the attractive and repulsive forces between nuclear particles**
- ✔ **the iter project**

A balance of forces

The nucleus of an atom is made up of protons and neutrons. This tells us something important – protons and neutrons are happy to stick together. There must be an attractive force that holds protons and neutrons together, and that can even hold two protons together despite the fact that they repel each other because of their positive charges. This force is called the **strong nuclear force**.

The strong nuclear force has a short range. It only acts when two nucleons (protons or neutrons) are very close together. In a nucleus, the particles are separated by just the right distance so that the strong nuclear force is balanced by the electrical (electrostatic) force.

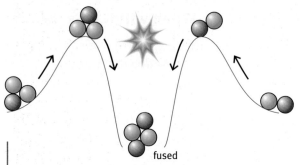

Pushing two hydrogen nuclei together. The 'hills' represent the repulsive force between them. The deep 'valley' represents the stable state they reach when they fuse together.

$$^{2}_{1}H + ^{3}_{1}H \rightarrow ^{4}_{2}He + ^{1}_{0}n$$

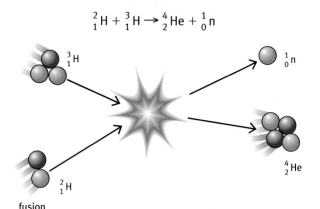

fusion
Fusion of two hydrogen isotopes gives helium.

Joining up

In the process of **nuclear fusion**, the nuclei of two hydrogen atoms join together and energy is released. The diagram on the right shows one way in which this happens. Note that the nuclei are of two different isotopes of hydrogen – both are hydrogen, because they have just one proton in the nucleus.

Picture bringing two atoms close together: their nuclei repel each other, because of the electrical (electrostatic) force. They will not fuse together. Push hard enough: they come close enough for the attractive force to take over, and the nuclei fuse. Energy is released.

You would have to do a lot of work to push two nuclei together, but you would get a lot more energy out when they fused.

Energy is released by a small amount of mass being converted to energy. The energy can be calculated from Einstein's equation $E = mc^2$.

It may seem strange that you can get energy from fission and fusion. It happens because the iron nucleus is the most stable, so nuclei lighter than iron can be joined to release energy and heavier nuclei can be split to release energy.

The quest for fusion power stations

If we could fuse the hydrogen nuclei from water to give nuclei of helium, which is an inert gas, we would have plenty of fuel (water) and the process would not produce as much nuclear waste.

Over the past 70 years there has been a lot of research. Scientists can give hydrogen nuclei enough energy to overcome the repulsive force. The problem is how to control the reaction and keeping it going?

When hydrogen isotopes are heated enough they lose their electrons and form a **plasma**. This is kept from touching the sides of the container by using magnetic fields. The JET project at Culham in the UK has researched fusion for many years. As yet no reactor has produced more energy than it used.

The ITER project

This is a joint project between China, the European Atomic Energy Community, India, Japan, Korea, Russia, and the U.S.A. ITER means 'the way' in Latin. Fusion research is very expensive so these countries have joined together to build a research reactor in France. Construction has begun. It will take 10 years to build, and be used for research for 20 years. ITER will investigate how plasmas behave during the hydrogen fusion reaction at 150 million °C. The goal is to be able to build a nuclear fusion power station.

You can think of fusion as the opposite of nuclear fission, the process used in nuclear power stations.

The H bomb

Hydrogen bombs, which fuse hydrogen, release hundreds of times more energy than atomic (fission) bombs. They are triggered using an atomic bomb to compress the hydrogen so that it fuses.

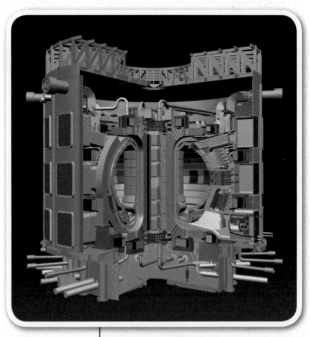

The planned ITER fusion reactor. Fusion will take place in the doughnut-shaped hole.

Questions

1 a What attractive force acts between particles in the nucleus? What repulsive force?
  b Which of these two forces has the greater range?

2 The Sun contains 1% oxygen nuclei. There are eight protons in an oxygen nucleus. Explain why these are less likely to fuse together than hydrogen nuclei.

3 Two hydrogen-2 nuclei can fuse to give a helium-3 nucleus.
  a Write a nuclear equation for the reaction.
  b What is the particle that is left over?

4 What are the advantages of countries working together on the ITER project?

5 ITER is very expensive. Write a letter to persuade the Government either:
  a to stay part of ITER or
  b to leave ITER to save money.

Key words
✓ **nuclear fusion**
✓ **strong nuclear force**

Science Explanations

Our understanding of radioactivity and the structure of the atom has enabled many applications, such as nuclear power stations and cancer treatment, to be developed. Knowledge of the way ionising radiation behaves is essential for working safely and making good risk assessments.

You should know:

- why some materials are radioactive and emit ionising radiation all the time
- how ionising radiation can damage living cells
- that atoms have shells of electrons and a nucleus made of protons and neutrons
- about the alpha scattering experiment and how it showed that the atom has a small, massive, positively charged nucleus
- that all the atoms of an element have the same number of protons
- that isotopes are atoms of the same element with different numbers of neutrons
- how the nucleus changes in radioactive decay
- how to complete nuclear equations for radioactive decay
- what alpha and beta particles and gamma radiation are, and their different properties
- that there is background radiation all around us, mostly from natural sources
- what radiation dose measures, and what factors affect it
- the difference between contamination and irradiation
- how to interpret data on risk related to radiation dose
- that radioactive materials randomly emit ionising radiation all the time and that the rate of decay cannot be changed by physical or chemical changes
- that the activity of a radioactive source decreases over time
- what is meant by the half-life of a radioactive isotope
- that radioactive isotopes have a wide range of half-life values
- how to do calculations involving half-life
- about uses of ionising radiation from radioactive materials
- about nuclear fission and energy released from the nucleus
- how nuclear power stations use the fission process to produce energy, including how the chain reaction is controlled, and about the nuclear waste produced
- about the three categories of radioactive waste, and the different methods of disposal
- that protons and neutrons in a nucleus are held together by the strong force, which acts against the electrical repulsive force between protons
- that hydrogen nuclei can fuse together to form helium if they are brought close enough together and this releases energy
- how to use Einstein's equation, $E=mc^2$, to calculate the energy released.

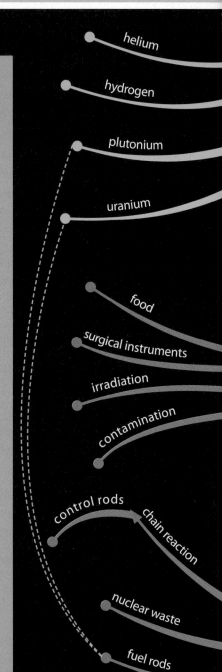

helium

hydrogen

plutonium

uranium

food

surgical instruments

irradiation

contamination

control rods

chain reaction

nuclear waste

fuel rods

RADIOACTIVE MATERIALS

E=mc²

isotopes

strong force

radioactive decay

nuclei

alpha particle scattering

energy released

half-life

reducing risk

environmental impact

fusion

fission

alpha and beta particles and gamma radiation

nuclear equations

exponential decay

mean

range

benefits

costs

penetration

structure of the atom

using equations

calculations using data

assessing risk

charge and nature

maths skills

who takes risk?

who benefits?

consequences of science

why are they radioactive?

health

risk

benefit

official regulations

treating cancer

as a tracer

decisions

sterilisation

uses

questions that science cannot answer

damage to living cells

background radiation

using and handling them safely

factors and outcomes

dose in sieverts

ITER project

ionising radiation

correlation

nuclear power stations

fission

fusion

energy from the nucleus

cause

Ideas about Science

In addition to developing an understanding of radioactive materials, it is important to appreciate the risks involved and how we make decisions about using science and technology. When considering risk, you should be able to:

- explain that nothing is completely safe. Everything we do has a certain risk. Background radiation is all around us, so there is always a risk of our cells being harmed by ionising radiation. But if the dose is low the risk is very small.
- list some of the uses of radioactive materials and the risks arising from them, both to people working with radioactive sources and to the environment.
- describe some of the ways that we reduce these risks.
- use data to compare and discuss different risks. Compare the risks of living in a high-radon area with risks of other activities, for example, smoking tobacco.
- discuss decisions involving risk. To decide whether to build a type of nuclear power station you would need to take account of the chance of contaminating the environment and how serious that would be.

- identify risks and benefits to individuals and groups. Many medical treatments make use of radioactive isotopes. The risk and the benefit to the patient must be weighed up.
- take into account, in making decisions, who benefits and who takes the risks, for example, when deciding whether to build a nuclear power station.
- suggest benefits of activities known to have risk, for example, a scan that involves injecting a radioactive tracer into the body.
- suggest reasons why people are willing (or reluctant) to take a risk. For example, people who are ill may choose to have treatment in the hope that they will be cured. Some people may refuse treatment if doctors say they must have it, as they prefer to choose for themselves.

In making decisions about science and technology, you should be able to:

- identify the groups affected by a decision, and the main benefits and costs for each group, for example, when deciding on a location for a waste-disposal site.
- explain that different decisions may be made depending on social and economic factors. Nuclear power stations are built in remote areas where they will affect fewer people. Countries with no other resources for generating electricity may choose to build nuclear power stations.
- explain that there is official regulation of some areas of research and application of knowledge. Countries that have nuclear power stations keep account of all the radioactive waste, as this could be processed to produce nuclear weapons.
- distinguish questions that can be answered using a scientific approach from those that cannot, for example, 'Shall we build a nuclear power station at this site?' If there is no scientific reason why not, the final decision still depends on what society wants to do.

Review Questions

1 Complete the sentences about radioactivity. Use key words from the module.

_____ radiation is produced by radioactive _____ . The radiation is produced when an _____ nucleus _____ . The three types of radiation produced are: _____, which is made up of two protons and two neutrons, _____, which is a high-energy electron, and _____, which is an electromagnetic radiation.

2 a What did the back-scattering alpha particles in the Rutherford scattering experiment show about atoms?

b How is the back-scattering explained by your answer to part a?

3 An isotope has a half-life of 74 days. Its activity is measured at 10 000 decays per second. It emits alpha radiation.

a What will its activity be after 133 days?

b How long will it take for the activity to reach 625 decays per second?

c For each example below, explain why the isotope would not be suitable and suggest an isotope from the module that would be suitable.

  i Measuring the age of rocks.

  ii A radioactive tracer in the body.

4 What is the difference between nuclear fission and nuclear fusion? Give examples of each with a nuclear equation.

5 The table shows some of the radioactive isotopes that are used in a range of applications.

Isotope	Radiation emitted	Half-life
americium-241	alpha	430 years
carbon-14	beta	5700 years
cobalt-60	gamma	5 years
iodine-123	gamma	13 hours
iodine-131	beta	8 days
strontium-90	beta	29 years
uranium-235	alpha	700 million years

For each application listed below choose the isotope you think most suitable and justify your answer, referring to both the half-life and the radiation emitted.

a calculating the age of rocks

b dating an ancient leather belt

c monitoring the thickness of paper in a factory

d monitoring uptake of iodine by the thyroid

e detecting smoke

f sterilising medical equipment

6 A nuclear power station uses the energy released by **nuclear fission** to generate electricity. The nuclear reactor is designed to control the **chain reaction**.

a Draw diagrams to explain what is meant by **nuclear fission** and a **chain reaction**.

b Describe the design features of the reactor that are used to control the chain reaction.

P7 Studying the Universe

Why study the Universe?

Physics offers an important way of looking at the world. Studying the whole Universe helps us know *where* the Earth is and *when* it is too because, it turns out, the Universe has a history. Though it may seem odd, explaining what happens in stars requires an understanding of matter at the microscopic scale, right down to the smallest subatomic particles. Everything in the physical world is made from a few basic building blocks.

What you already know

- The Solar System includes planets, asteroids, minor planets, and comets, all orbiting the Sun.

- The energy of the Sun is the result of nuclear fusion in the core of the star.

- The Universe began with a 'big bang', and after 14 thousand million years it is still expanding.

Find out about

- what we can see in the night sky

- how telescopes work and how they are used to map the Universe

- the life stories of stars

- how the astronomy community works together.

The Science

In the 1830s, the French philosopher Auguste Comte suggested that there were certain things that we could never know. As an example, he gave the chemical composition of the stars. By 1860, two years after his death, physicists had interpreted the spectrum of starlight and identified the elements present. In this module, you will look at how physicists have gradually extended our understanding of stars and galaxies.

Ideas about Science

How can scientists be sure? Scientists make careful observations of the Solar System and Universe. They use their imaginations to explain the data. Then they test their ideas by sharing them with the wider science community.

Topic 1: Naked-eye astronomy

This sailor is demonstrating the use of an astrolabe, an astronomical instrument invented over 2000 years ago. The instrument measures the angle of a star above the horizon. It was used for navigation, astronomy, astrology, and telling the time.

Astronomy is the oldest science in the world. Ancient civilisations – for example, the Chinese, Babylonians, Egyptians, Greeks, and Mayans – practised naked-eye astronomy long before the invention of telescopes in the 17th century. They built costly observatories for both practical and religious reasons.

Everything beyond the Earth moves across the sky, from horizon to horizon. Calendars and clocks were based on cycles in these movements: day and night, the phases of the Moon, and seasonal changes in the Sun's path. Long-distance travellers, on sea and on land, navigated using the positions of familiar stars.

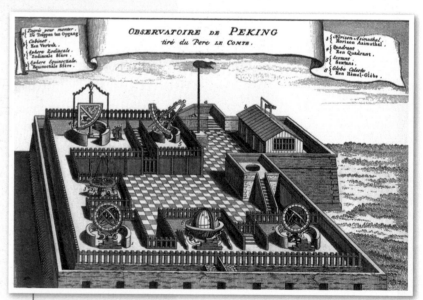

The Chinese Imperial Observatory was built in 1442. Instruments similar to astrolabes were used to measure the positions of objects in the sky.

Day-time astronomy

You do not have to stay up all night to make valid astronomical observations. You can see the Sun cross the sky every day from East to West, moving at a steady rate. That is an observation that any scientific theory of the Universe must account for. You may also have noticed that the Moon follows a similar path, sometimes by day and sometimes by night.

Around the pole

The stars also move across the night sky. Their movement is imperceptible, but it is revealed by long-exposure photography. The photo in the margin below shows that the stars appear to rotate about a point in the sky directly above one of the Earth's poles.

Eclipses

Sometimes it is rare and unusual events that reveal something important. When the Sun passes behind the Moon in a total eclipse, you can see the Sun's gaseous corona.

The Sun sets – a time-lapse photograph.

During a total eclipse of the Sun, the outer atmosphere, or corona, becomes visible. Its appearance changes from one eclipse to the next.

This photograph shows the motion of the stars across the sky. The exposure time was 10.5 hours.

Find out about

- observations with the naked eye
- phases of the Moon
- how stars change position during the night
- the difference between sidereal and solar days
- how astonomers describe positions in the sky

The Moon is shown here at intervals of three days. The Sun is off to the right. Notice that it is the half of the Moon facing the Sun that is lit up.

Key words

- phases of the Moon
- constellation

The spinning observatory

The Sun and Moon move across the sky in similar but slightly different ways:

- The Sun appears to travel across the sky once every 24 hours (on average).
- The Moon moves very slightly slower, reappearing every 24 hours and 49 minutes.

Arctic Sun. A series of photographs over 24 hours shows the Sun's position in the sky each hour during a single day in the Arctic summer.

People are, of course, deceived by their senses. The Sun is not moving round the Earth. It is the Earth that is spinning on its axis. That is why the Sun rises and sets every day, and why we experience day and night.

The situation with the Moon is more complex. The spinning of the Earth makes the Moon cross the sky. But the Moon is also slowly orbiting the Earth, from West to East. One complete orbit takes about 28 days.

Even without taking into account the fact that the Earth orbits the Sun, you can use these ideas to explain the changing **phases of the Moon**.

The Moon's phases

At any time, half of the Moon is lit up by the Sun's rays, just like the Earth. The view from the Earth depends on where the Moon is around its orbit.

- When the Moon is on the opposite side of the Earth to the Sun, an observer on the Earth can see the whole of its illuminated side. This is a full Moon.
- When the Moon is in the direction of the Sun, the side that is in darkness faces the Earth. This is a new Moon.

The phase of the Moon changes as it orbits the Earth.

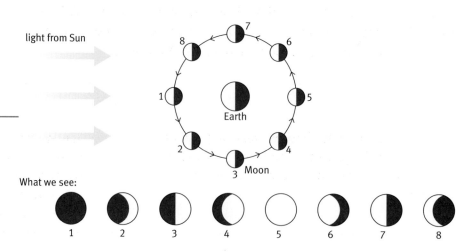

Mapping the sky: constellations

The stars are always seen in the same patterns. Most civilisations have identified 'pictures' in the stars and given them names. The stars within each **constellation** (group of stars) are usually separated by huge distances and have no connection with one another. Different civilisations have 'joined the dots' differently, so they have different constellations. Many of the constellation names we use today originated with the ancient Greeks. Astrologers use the movements of planets across constellations to tell fortunes, but this is superstition and there is no scientific evidence that it works.

The Southern Cross, photographed from Cuba.

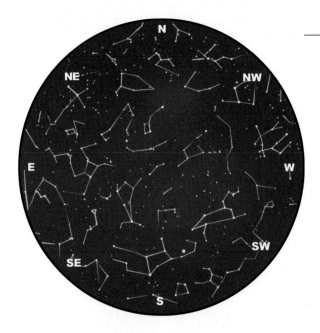

Star charts from newspapers and websites show which stars and planets can be seen at a given time and date. First-magnitude stars are the brightest and can most easily be seen.

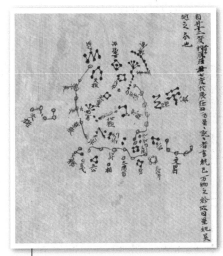

This Chinese star map was probably drawn in the seventh century. The stars are joined in much smaller groups than in European maps.

Seasonal skies

Some constellations seen on a winter night are different from those of a summer night. This is because the Earth travels halfway round its orbit in six months. You see the stars that are in the opposite direction to the Sun, so after six months you will see the opposite half of the sky.

Each day, a star seen from Earth will rise four minutes earlier. After six months, those extra minutes add to twelve hours, so that a star that is rising at dusk in June will be setting at dusk in December.

Questions

1 Draw a diagram to show the relative positions of the Earth, Sun, and Moon when the Moon is at first quarter (half-illuminated as seen from Earth).

2 Imagine that the Earth suddenly starts to spin in the opposite direction. What difference would this make to:
   a the path of the Sun and Moon across the sky?
   b the Moon's phases?

The spinning, orbiting observatory

Earth-bound observers see the sky from a rotating planet. That is why the stars appear to move across the night sky. Their apparent motion is slightly different from the Sun's.

• The stars appear to travel across the sky once every 23 hours 56 minutes.

That is four minutes less than the time taken by the Sun. The difference arises from the fact that the Earth orbits the Sun, once every year.

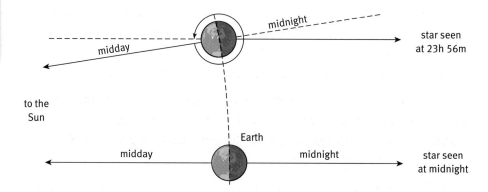

Imagine looking up at a bright star in the sky. 23 hours 56 minutes later, it is back in the same position. This tells you that the Earth must have turned through 360° in this time, and you are facing in the same direction in space.

Repeat the above observation, this time looking at the Sun. After 23 hours 56 minutes, the Earth has turned through 360°, but the Sun has not quite reached the same position in the sky. The diagram shows that, in the course of a day, the Earth has moved a short distance around its orbit. Now it must turn a little more (4 minutes' worth) for the Sun to appear in the same direction as the day before.

Days are measured by the Sun. The average time it takes to cross the sky is 24 hours, and this is called a **solar day**. We could choose to set our clocks by the stars (although this would be very inconvenient). Then a day would last 23 hours 56 minutes; this is called a **sidereal day** ('sidereal' means 'related to the stars').

The celestial sphere

Long ago people described the stars as though they were lights on the inside of a spinning bowl, with the observer on the Earth at the centre. A better description extends the bowl to be a complete sphere. This **celestial sphere** has an axis running from the Pole Star through the axis of the Earth. The celestial equator is an extension of the Earth's equator.

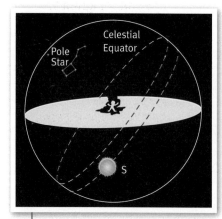

You can think of the Sun and stars as fixed. We view them from a spinning, orbiting planet.

Key words

✓ solar day
✓ sidereal day

The celestial sphere is an imaginary sphere with the Earth at its centre and the stars fixed to the surface of the sphere.

How can we locate objects in the sky?

By international agreement, the sky is divided up into regions named after constellations. Astronomers use these to describe positions in the sky.

To give positions more precisely, astronomers use angles.

Imagine standing in a field, looking at a star. Two angles are needed to give its position:

- Start by pointing at the horizon, due North of where you are. Turn westwards through an angle until you are pointing at the horizon directly below the star. That gives you the first angle.
- Now move your arm upwards through an angle, until you are pointing directly at the star. That gives you the second angle.

But the stars move across the sky every 24 hours, and are in different directions depending on where on the Earth's surface you are standing. So astronomers use a system that will work wherever you are and whatever the time of day and year.

The equatorial coordinate system uses a reference point in the sky (this point is called the vernal equinox point). The position is defined using two angles – **right ascension** and **declination**. The diagram below shows how these are used.

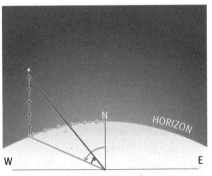

Two angles describe the position of a star.

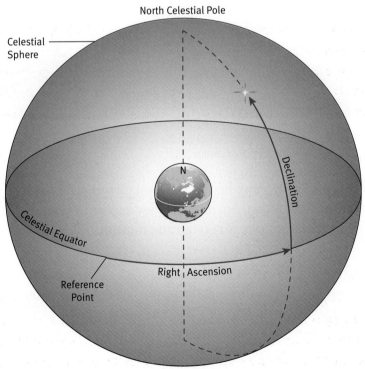

Right ascension measures the angle east from the vernal equinox point. The angle is commonly measured in hours, minutes, and seconds. Declination measures the angle of the star above or below the celestial equator. The angle is usually measured in degrees and minutes.

Questions

3 a If the Earth orbited the Sun more quickly – in, say, 30 days – would the difference between sidereal and solar days be greater or less?

   b Work out the time difference between sidereal and solar days.

4 Why would it be 'inconvenient' if we set our clocks according to sidereal time?

5 a Draw a diagram to explain why you see some different constellations in winter and summer.

   b Use your diagram to explain why there are some stars that can never be seen from the UK, but that can be seen from places in the southern hemisphere.

6 Suggest how measuring the angles of stars above the horizon could help with navigation.

Find out about

- ✓ **how planets move against the star background**
- ✓ **how planets' motion can be explained**

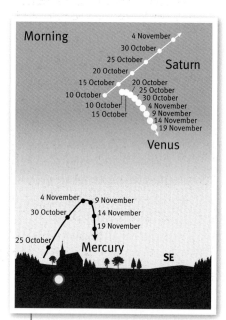

The pattern of movements of planets in the sky is different from that of stars. This diagram shows the pattern of movement of three planets just before sunrise.

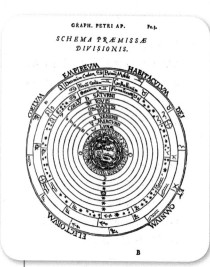

This geocentric (Earth-centred) map was published in Germany in 1524.

Heavenly wanderers

Five **planets** can be seen with the naked eye from Earth – Mercury, Venus, Mars, Jupiter, and Saturn. These were recognised as different from stars long, long ago because they appear to move, very slowly, night by night, against the background of 'fixed stars'. The diagram on the left shows the changing positions in the sky of three planets at dawn over a few weeks in one recent year. Any scientific theory of the planets must be able to explain their observed motion.

Mercury and Venus are only ever seen at dawn or dusk, fairly close to the horizon near the setting or rising Sun.

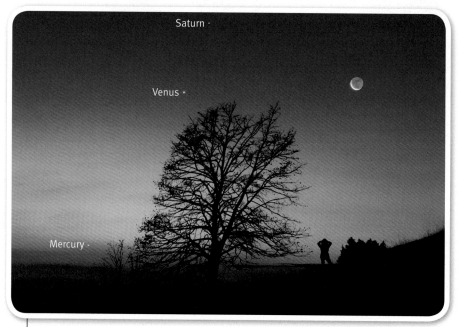

The planets appear as bright objects in the sky. Here Mercury, Venus, and Saturn are visible as well as the crescent Moon.

The planets generally move in an East–West direction along a similar path to the Sun and Moon. But at times they appear to slow down and go into reverse; this is known as **retrograde motion**.

What is at the centre?

Before about 1600, maps showing the layout of the Universe placed the Earth at the centre, with the Sun, Moon, planets, and stars orbiting around it. But to explain the planets' retrograde motion, their **orbits** would need to be very complicated.

In 1543, the Polish astronomer Copernicus suggested that Earth, along with the five known planets, orbited the Sun. This idea was rejected by many people at the time, but its success at explaining many observations meant that eventually it became accepted.

Planets' motion explained

In our present-day model of the **Solar System**, Earth is a planet.
The planets orbit the Sun, and their orbits all lie in approximately the
same plane. Each planet takes a different amount of time to orbit the
Sun. The planets furthest from the Sun take the longest time.

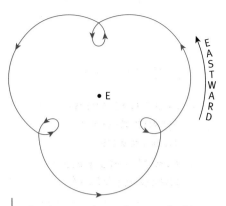

This diagram shows the sort of orbit a
planet would need to have if it orbited
the Earth in a way consistent with the
observations.

Earth is one of eight major planets that orbit the Sun. Pluto is a
dwarf planet; its orbit is tilted and is a different shape. (In this
picture, the planets and Sun are not drawn to scale.)

To explain retrograde motion, recall that both the Earth and the planets
are orbiting the Sun. An observer looking towards Mars sees it against a
backdrop of the fixed stars. Its position against this backdrop depends
on where the Earth and Mars are in their orbits.

Key words

- ✓ **planets**
- ✓ **retrograde motion**
- ✓ **orbits**
- ✓ **Solar System**

Questions

1 Mercury is the closest
planet to the Sun. It is only
ever seen at dawn or dusk,
close to the Sun in the sky.
Draw a diagram to
explain why.

2 Suggest reasons why
people rejected the
Sun-centred idea and
why it took a long time
to be accepted.

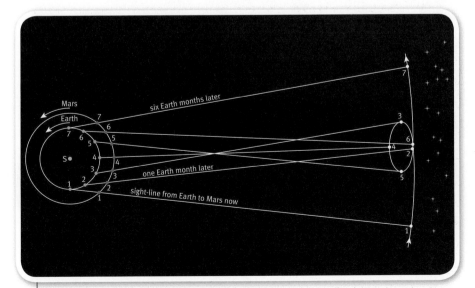

From months 1 to 3, Mars appears to move forwards. Then, for two months, it goes
into reverse before moving forwards again.

Find out about

- ✔ why solar and lunar eclipses happen
- ✔ angular size
- ✔ the effect of the Moon's orbital tilt

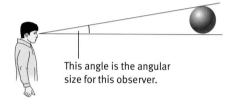

This angle is the angular size for this observer.

Angular size is the angle between two lines drawn from an observer to each side of an object. It depends on the object's actual size and its distance away.

Key words

- ✔ solar eclipse
- ✔ lunar eclipse
- ✔ angular size

Eclipses involve both the Sun and the Moon.

- In a **solar eclipse**, the Moon blocks the Sun's light.
- In a **lunar eclipse**, the Moon moves into the Earth's shadow.

Astronomers can predict when an eclipse of the Sun will occur. A solar eclipse happens just a few times each year. And a total eclipse at any particular point on the Earth is a rare event.

The predictability of eclipses shows that they must be related to the regular motions of the Sun and Moon. Their rarity suggests that some special circumstances must arise if one is to occur.

The first way to explain a solar eclipse is to think of the Sun and Moon and their apparent motion across the sky. The Sun moves slightly faster across the sky than the Moon, and its path may take it behind the Moon. For us to see a total eclipse, the Sun must be travelling across the sky at the same height as the Moon. Any higher or lower and it will not be perfectly eclipsed.

The fact that the Moon precisely blocks the Sun is probably a coincidence. The Sun is 400 times the diameter of the Moon, and it is 400 times as far away. So, by coincidence, they both have the same **angular size** (about 0.5°).

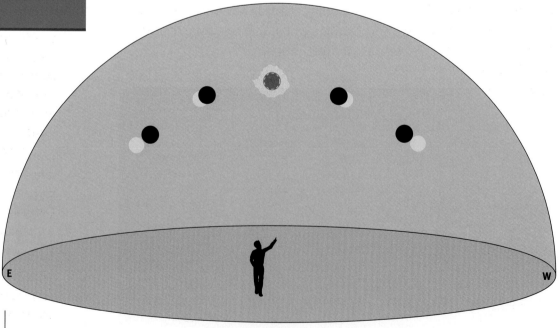

Provided the Sun's path across the sky matches the Moon's, a total eclipse may be seen.

Umbra and penumbra

The diagram below shows a different way of explaining eclipses, both solar and lunar. Both the Earth and the Moon have shadows – areas where they block sunlight. Because the Sun is an extended source of light, these shadows do not have hard edges. There is a region of total darkness (the umbra) fringed by a region of partial darkness (the penumbra). The Earth's shadow is much bigger than the Moon's.

• When the Moon's umbra touches the surface of the Earth, a solar eclipse is seen from inside the area of contact.

• When the Moon passes into the Earth's umbra, a lunar eclipse is seen.

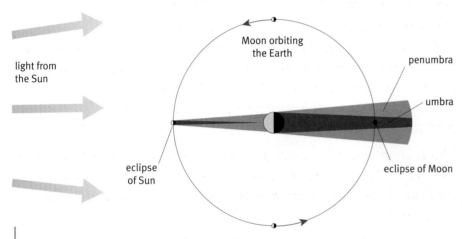

The umbra and penumbra for an eclipse of the Sun and an eclipse of the Moon.

Why the rarity?

The Moon orbits the Earth once a month, so you might expect to see a lunar eclipse every month, followed by a solar eclipse two weeks later. You do not – eclipses are much rarer than this. The reason is that the Moon's orbit is tilted relative to the plane of the Earth's orbit by about 5°. Usually the Earth, Sun, and Moon are not in a line so no eclipse occurs.

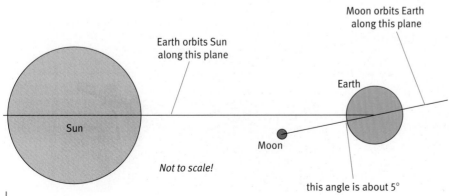

The Moon's orbit is tilted relative to the plane of the Earth's orbit around the Sun. The effect is exaggerated here.

Questions

1 At times, the Moon's orbit takes it further from Earth so that it looks smaller in the sky. If it is in front of the Sun, the result is that we see a ring of bright sunlight around the black disc of the Moon. This is an annular eclipse. Construct a diagram like the one on this page to show why this happens.

2 What is the phase of the Moon at the time of:
   a a solar eclipse?
   b a lunar eclipse?

3 Explain why a person on Earth is more likely to see a lunar eclipse than a solar eclipse.

Eclipse trips

Today, solar eclipses are big business. Thousands of people select their holiday dates to coincide with an eclipse. Tour operators organise plane-loads of eclipse spotters, and cruise liners sail along the track of the eclipse. Guest astronomers give lectures to interested audiences. And, provided the clouds hold off, hundreds of thousands of satisfied customers will get a view of a spectacular natural phenomenon.

The moment of total eclipse. For a few tens of seconds, the Moon blocks the Sun's bright disc and the solar corona (the Sun's hot atmosphere) is visible. Photograph by Fred Espenak.

Scientific expeditions

For centuries, astronomers have travelled to watch eclipses for scientific purposes. They have helped us to learn about the dimensions of the Solar System, and about the Sun and Moon. Take the question of the corona. For a long time, scientists had been unable to agree whether the corona was actually part of the Sun, or a halo of gas around the Moon, illuminated by sunlight during an eclipse.

The picture on the right shows a scientific expedition that travelled to India to observe and record the total solar eclipse of 12 December 1871. They took photographs from which it was possible, for the first time, to develop a scientific description of the Sun's corona (outer atmosphere).

A scientific purpose?

But is it worth studying eclipses today? Are there good scientific reasons to take tonnes of scientific equipment off to some distant land? One person who thinks so is Fred Espenak, an astrophysicist at NASA's Goddard Space Flight Centre. His interest is in the atmospheres of planets, moons, and the Sun.

Preparing to observe a solar eclipse in 1871.

Fred uses an infrared spectrometer to examine radiation coming from planetary atmospheres. A spectrometer is a device that splits radiation into its different frequencies. For example, a prism splits light into the colours of the spectrum from red (lowest frequency) to violet (highest frequency). By studying the frequencies that are present, it is possible to deduce the chemical composition of the source of the radiation. One of Fred's experiments, to measure atmospheric flow, was carried out on the Space Shuttle.

A young Fred Espenak, preparing to observe an eclipse in 1983. He is now a veteran of over 20 eclipse expeditions.

A solar eclipse lets Earth-bound observers see the Sun's corona. This is a mysterious part of the Sun, extending far out into space. The mystery is its temperature. We see the surface of the Sun, which is hot, at about 5500°C, but the corona is far hotter – perhaps 1.5 million degrees. Measurements during eclipses may help to explain how this thin gas become heated to such a high temperature.

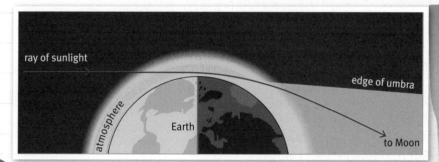

ray of sunlight
edge of umbra
atmosphere
Earth
to Moon

During a lunar eclipse, light from the Sun is refracted as it passes through the Earth's atmosphere, and lights up the Moon.

And lunar eclipses? Fred's main interest is in examining the light that reaches the Moon through the Earth's atmosphere at this time. The quality of the light can be a good guide to the state of the Earth's atmosphere, indicating pollution from such causes as forest fires and volcanoes.

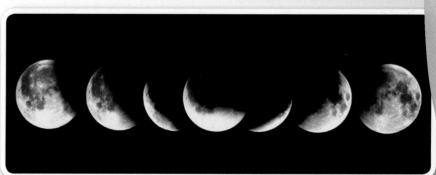

A composite image of the Moon moving in and out of eclipse. The central image shows the Moon lit up by sunlight that has been refracted through the Earth's atmosphere. Photograph by Fred Espenak.

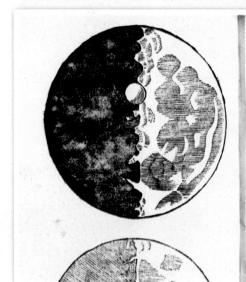 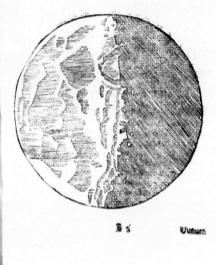

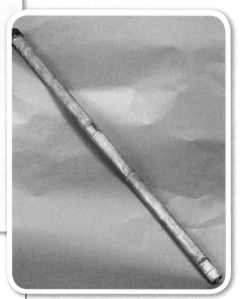

Galileo made these sketches of the Moon in 1610, and saw that the edge of the shadow between the Moon's light and dark sides was sometimes irregular (top) and sometimes smooth (bottom). He deduced that this was due to mountains on the Moon, challenging the existing worldview that said the heavens were perfect and unchanging.

This telescope is a replica of one made by Galileo. It uses lenses to focus light. His best telescopes had a magnification of about 30 times.

In the autumn of 1609, Galileo made his first observations of the Moon using a telescope. He was not the first person to use a telescope to look at the night sky. But the observations Galileo made, and his interpretation of them, had repercussions down the centuries. He changed the way people thought about the Universe.

Everything that we know about the Universe beyond the Solar System comes from using telescopes. By studying electromagnetic radiation from very distant objects, astronomers can discover a surprising amount of information about things that are much too far away to visit.

Seeing the light

The first telescopes used visible light. These are known as **optical telescopes**. Some are designed for people to look through, but most modern optical telescopes record the image using electronic detectors.

Knighton Observatory in mid-Wales is part of the Spaceguard Foundation. It keeps an eye out for comets and asteroids that might collide with Earth.

Greenwich Observatory in south-east London was built by the British Navy in the 17th century. They had suffered defeat at the hands of the Dutch because the enemy were able to navigate better. Dutch astronomy was in advance of the English, giving them better star charts of the night sky, which were used to work out the ship's position.

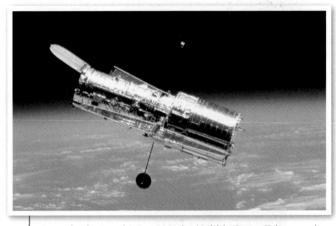

Launched into orbit in 1990, the Hubble Space Telescope has given us a more detailed view of the Universe than ever before.

This reflecting telescope is at the Calar Alto Observatory, over 2000 m above sea level in southern Spain. Light passes through the ring (diameter 3.5 m) and reflects off the curved, shiny mirror at the back.

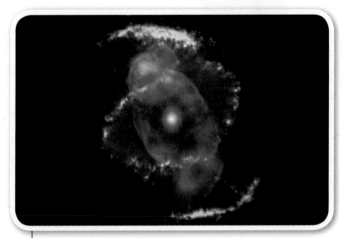

The Hubble Space Telescope made this image in 2004. It shows the Cat's Eye nebula, in which thin layers of hot gas have been thrown off the surface of a dying star.

Key word

- ✓ **optical telescopes**

Find out about

✓ how telescopes can reveal detail that is not seen with the naked eye

Key words

✓ aperture
✓ telescope
✓ image

A telescope makes distant objects appear larger and closer, so that more detail can be seen. Also, it has a bigger **aperture** than a human eye. This means it collects more radiation, allowing fainter objects to be studied.

Modern telescopes can produce lasting images using electronic detectors or photographic film. Also, images of faint objects can be made by collecting radiation over a long time.

Beyond the visible

Since the mid-20th century, astronomers have developed telescopes to study radiation in all parts of the electromagnetic spectrum. Each requires a suitable detector and a means of focusing the radiation. These telescopes have produced results that came as a complete surprise, leading to major changes in people's ideas about stars and the Universe.

Jocelyn Bell with part of one of the charts produced by the radio telescope, showing the trace produced by a pulsar.

Pulsars

In October 1967, Jocelyn Bell was working with Anthony Hewish in Cambridge. She was involved in making a survey of radio sources in the sky.

'Six or eight weeks after starting the survey I became aware that on occasions there was a bit of "scruff" on the records, which did not look exactly like man-made interference. Furthermore I realised that this scruff had been seen before on the same part of the records – from the same patch of sky.

Whatever the source was, we decided that it deserved closer inspection, and that this would involve making faster chart recordings. As the chart flowed under the pen, I could see that the signal was a series of pulses, and my suspicions that they were equally spaced were confirmed as soon as I got the chart off the recorder. They were 1.3 seconds apart.'

Jocelyn Bell and Anthony Hewish had discovered pulsars. These are distant objects that send out radio waves that vary with an extremely regular pulse.

Jocelyn Bell and Anthony Hewish used a radio telescope. The wires in the field form an aerial that detects radio waves. There are 1000 posts spaced over 4.5 acres.

The Effelsberg radio telescope in Germany. With a diameter of 100 m, it gathers radio waves from distant objects in the Universe, including galaxies other than our own. The dish is scanned across the sky to generate an image.

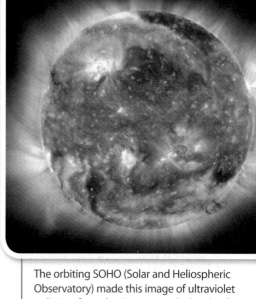

The orbiting SOHO (Solar and Heliospheric Observatory) made this image of ultraviolet radiation from the Sun. It reveals details of the hot atmosphere that traces the shape of the Sun's magnetic field.

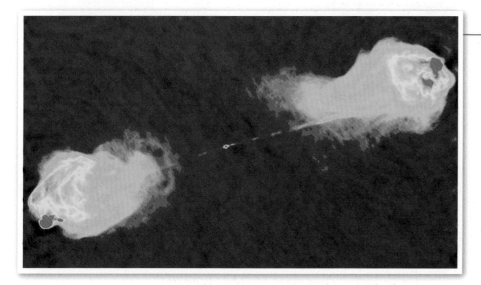

This image shows radio emission from a galaxy called Cygnus A, one of the brightest radio objects in the sky. With an optical telescope, all you observe is the tiny central dot, which is a distant galaxy. The radio waves come from clouds stretching a quarter of a million light-years either side of the galaxy.

Infrared radiation is mostly absorbed by Earth's atmosphere. The orbiting Spitzer telescope observes infrared radiation from space. This infrared image was taken with the Spitzer telescope. The green indicates clouds of warm gas between stars. The small red spots are hot balls of gas on their way to becoming stars, and the darker patch on the left is the remains of an exploded star.

Questions

1 Different telescopes make use of different types of electromagnetic radiation. List the telescopes shown on these two pages, together with the radiation that each gathers.

2 Telescopes 'make things visible that cannot be seen with the naked eye'. Is that true for all of the telescopes shown here?

3 Look at the image of the Moon on page 208 and think about how the Moon appears to the naked eye when you see it in the night sky. What extra features has Galileo been able to identify by using a telescope?

Find out about

- ✔ **how to make an image using a pinhole camera**
- ✔ **how to draw diagrams of light rays to explain image formation**

Making an image

A telescope makes an image of a distant object that is clearer and more detailed than the view with the naked eye. To understand how telescopes work, you need to learn how images are formed.

The simplest way to make an image uses just a small pinhole in a sheet of card. Light that is scattered, or given out, by an object passes through the pinhole and makes an image on a screen. This is a **real image**. A real image can be recorded using a light-sensitive electronic detector or photographic film. You can see a real image on a screen without having to look towards the object.

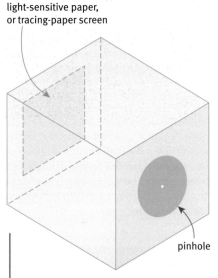

light-sensitive paper, or tracing-paper screen

pinhole

You can use a pinhole camera, made from a box with a pinhole at one end, to take photographs using light-sensitive paper or film. Alternatively, use a tracing-paper screen so that you can see the image from outside.

This picture was taken with a pinhole camera.

Ray diagram

A **ray diagram** helps explain how a pinhole produces a real image on a screen.

- Straight lines with arrows show the direction of light rays travelling from the object.

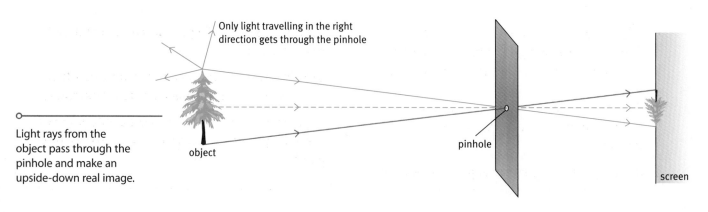

Only light travelling in the right direction gets through the pinhole

Light rays from the object pass through the pinhole and make an upside-down real image.

object

pinhole

screen

- Rays leave the object in all directions, but only those travelling towards the pinhole get through and light up the screen.
- As the pinhole is very small, each point on the object produces a tiny spot of light on the screen. Together, all these spots make up the image.

A ray diagram for a pinhole camera can help you predict and explain:
- why the image is upside down
- what happens to the image if the screen is moved away from the pinhole
- what happens if you enlarge the pinhole.

Pinhole telescope?

You can use a pinhole to make an image of the Sun. The image is faint because the aperture is small – only a small amount of light passes through the pinhole.

This long-exposure photograph was taken with a pinhole camera. It shows the path of the Sun over six months.

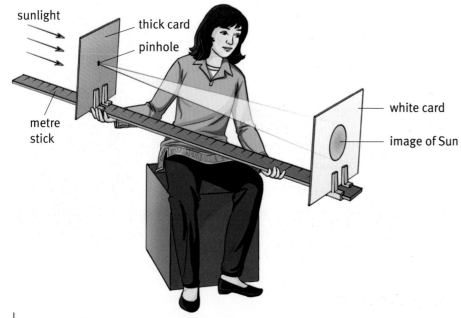

sunlight

thick card

pinhole

metre stick

white card

image of Sun

SAFETY: Face away from the Sun so that it shines over your shoulder. Do not look directly at the Sun.

Questions

1 a Draw two ray diagrams to explain what happens to the image in a pinhole camera if the screen is moved further away from the pinhole.
   b Predict what happens to the image if the distance between pinhole and screen is doubled.

2 By drawing a ray diagram, predict what you would see on the screen of a pinhole camera that had several pinholes.

3 Explain why a pinhole image of stars other than the Sun would not be much use to astronomers.

4 Suggest reasons why the pinhole photograph of the Sun's path above has some dark streaks and patches.

Find out about

✓ **how a converging lens makes a real image**
✓ **the focal length and power of a lens**
✓ **refraction**

The earliest and simplest telescopes used lenses. To understand these telescopes, you need first to understand what lenses do.

The focal length of a converging lens

The picture below shows how you can make a miniature image of a distant object using a single **converging lens**. If you position the screen correctly, you will see a small, inverted image of a distant scene.

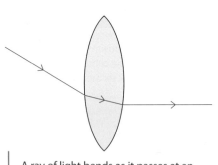

A ray of light bends as it passes at an angle from one material to another. This effect is called **refraction**.

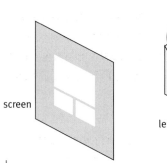

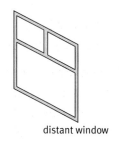

distant window

screen

lens in holder

Using a converging lens to make an image on a screen.

Rays of light enter the lens. Because of the lens shape they are refracted (they change direction), first on entering the lens and again on leaving.

A ray diagram shows this. A horizontal line passing through the centre of the lens is called the **principal axis**. Rays of light parallel to the axis are all refracted so that they meet at a point. This is why converging lenses are so called: they cause parallel rays of light to converge.

How a converging lens focuses parallel rays of light.

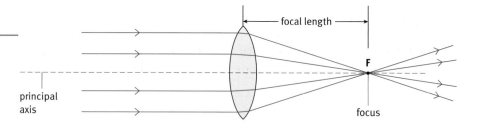

focal length

F

principal axis

focus

Lenses are cleverly designed. The surface must be curved in just the right way in order for the rays to meet at a point, the **focus** (F).

When early astronomers made their own telescopes, they had to grind lenses from blocks of glass. If the glass was uneven, or if the surface was not smooth or was of the wrong shape, the telescope would give a blurred, poorly focused image. It is said that much of Galileo's success was achieved because he made high-quality lenses so that he could see details that other observers could not.

The distance from the centre of the lens to the focus is called the **focal length** of the lens. The longer the focal length of a lens, the larger (actual physical size) will be the real image that the lens produces of a distant object.

Estimating the focal length

You can compare lenses simply by looking at them. For lenses of the same material:

* a lens with a long focal length has surfaces that are not very strongly curved
* a lens with a short focal length has surfaces that are more strongly curved.

To estimate the focal length of a converging lens, stand next to the wall on the opposite side of the room to the window. Hold up the lens and use it to focus an image of the window on the wall. Measure the distance from the lens to the wall – this will give you a good estimate of the focal length.

The power of a lens

A lens with a short focal length bends the rays of light more. Its **power** is greater. So short focal length equals high power, and long focal length equals low power. As one increases, the other decreases, they are inversely proportional.

Here is the equation used to calculate power when you know the focal length:

$$\text{power (in \textbf{dioptres}, \textbf{D})} = \frac{1}{\text{focal length (in metres)}}$$

So if a lens has a focal length of 0.5 m, its power is

$$\frac{1}{0.5} = 2 \text{ dioptres}$$

If you look at the reading glasses sold in chemists' shops, or at an optician's prescription, you will see the power of the lens quoted in dioptres.

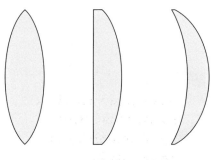

converging lenses

All converging lenses are fatter in the middle than at the edges.

Questions

1 The focal length of a lens is measured between which two points?

2 A lens has a focal length of 20 cm. What is its power?

3 A pair of reading glasses has lenses labelled +1.5D (D stands for dioptres). What is their focal length?

Key words
✔ **converging lens**
✔ **principal axis**
✔ **focus**
✔ **focal length**
✔ **power**
✔ **dioptres**

More about lenses

To astronomers, most objects of interest are so distant that they are 'at infinity'. When light reaches a lens from such a distant point, all the rays entering the lens are parallel. Parallel rays at an angle to the principal axis converge to one side of the principal focus.

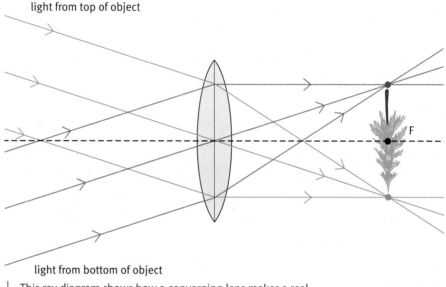

light from top of object

F

light from bottom of object

This ray diagram shows how a converging lens makes a real image of a distant object.

If you look through a converging lens at a very nearby object, you see an image that appears to be behind the lens. It is bigger than the object and the right way up. The lens is acting as a magnifying glass.

If you look through a diverging lens you see a virtual image that appears smaller than the object and the right way up. A diverging lens on its own cannot make a real image.

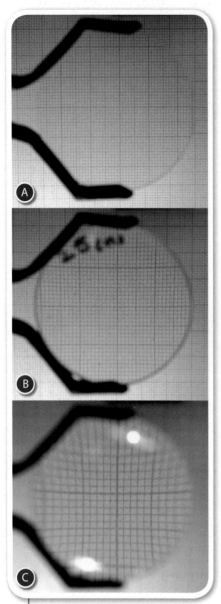

Three lenses with different focal lengths.

Questions

4 Look at the three lenses A, B, and C in the photograph. (The lenses are made from the same material.)
   a Put them in order starting with the one with the shortest focal length.
   b Which one has the greatest power?
   c How are their shapes different?

5 Suppose that you adapted a pinhole camera so that it had a large aperture and a converging lens instead of the pinhole. What differences would this make to the image of a distant object?

More about refraction

Light changes direction when it passes at an angle from one material to another. This is called **refraction**. To understand why refraction happens, it is helpful to look at the behaviour of water waves.

Waves can be studied using a ripple tank. Waves are created on the surface of water and a lamp projects an image of the waves on to a screen.

Waves in water of constant depth are equally spaced. This shows that waves do not slow down as they travel. The wave speed stays the same. As the wave travels, it loses energy (because of friction). Its amplitude gets less but not its speed.

Refraction

If waves cross a boundary from a deeper to a shallower region, they are closer together in the shallow region. The wavelength is smaller. This effect is called refraction. It happens because water waves travel slower in shallower water. The frequency (f) is the same in both regions, so the slower wave speed (v) means that the wavelength (λ) must be less (as $v = f\lambda$).

If the waves are travelling at an angle to the boundary between the two regions, their direction also changes. You can work out which way they will bend by thinking about which side of the wave gets slowed down first.

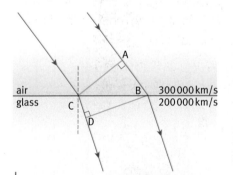

Light is refracted because its speed is different in the two media. Light travels the distance AB in air in the same time as it travels the distance CD in glass.

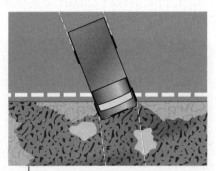

A model to help explain refraction. The truck goes more slowly on the muddy field. One wheel crosses the boundary first – so one side slows down before the other. This makes the truck change direction.

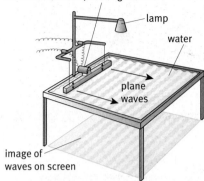

A ripple tank producing a steady stream of plane waves.

Plane waves travelling at constant speed.

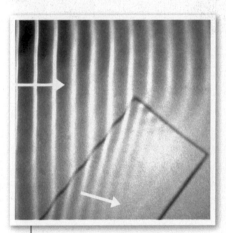

Refraction of water waves at a boundary between deep and shallow regions.

Question

6 When a ray of light passes from air into glass, it slows down.
   a Explain what happens to its frequency and to its wavelength.
   b Draw a diagram to show how the direction of the light changes as it passes at an angle from air into glass.

Key word
✓ **refraction**

Find out about

- ✔ **the lenses used in telescopes**
- ✔ **the magnification produced by a telescope**
- ✔ **how bigger apertures give brighter images**

A telescope that uses two converging lenses.

Just two lenses

Telescopes evolved from the converging lenses used for correcting poor eyesight. These were in use before 1300, though you had to be quite 'well off' to afford a pair of spectacles in those days. The lenses were biconvex; that is, they were convex (bulging outwards) on both sides. (The word 'lens' is Latin for 'lentil', which has the same shape.)

Such lenses work as magnifying glasses, producing an enlarged image when you look through them at a nearby object. For medieval scholars, whose eyesight began to fail in middle age, spectacles meant that they could go on working for another 20 or 30 years.

The Dutch inventor Hans Lippershey is credited with putting two converging lenses together to make a telescope. There is a story that in 1608, his children held up two lenses and noticed that the weathercock on a distant building looked bigger and closer. Lippershey tested their observation, and went on to offer his invention to the Dutch military.

In fact, Lippershey failed to get a patent on his device. Other 'inventors' challenged his claim, and the Dutch government decided that the principle of the telescope was too easy to copy, so that a patent could not be granted.

Telescopes rapidly became a fashionable item, sold widely across western Europe by travelling salesmen. That is how one came in to Galileo's hands in Padua, near Venice, in 1609.

Hans Lippershey and his children. Their play with lenses led to the invention of the telescope.

DIY telescope

A telescope using lenses to gather and focus the light is called a refracting telescope, or a **refractor**. You can make a telescope using almost any two converging lenses mounted in a line.

- The **eyepiece lens** is next to your eye. This is the stronger of the two lenses.
- The **objective lens** is nearer to the object you are observing. This is the weaker lens.

You will only get a clear image if the two lenses have the correct separation, and this depends on the distance to the object. You will need to adjust the separation so that you get a clear view. This is called **focusing**.

This is what you can see through the telescope shown on the opposite page. The image is upside down.

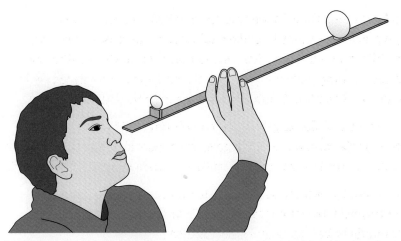

Looking through the eyepiece, you should see an **inverted** (upside-down) **image** of a distant object.

Key words
- ✓ **eyepiece lens**
- ✓ **objective lens**
- ✓ **focusing**
- ✓ **inverted image**

Ray diagram of a telescope

You can use what you have learned about lenses to explain how a telescope made of two converging lenses works. The diagram below shows what happens.

- The objective lens collects light from a distant object.
- Parallel rays of light enter the objective lens from a point on the distant object.
- Each set of parallel rays is focused by the objective lens, so a real image is formed.
- The eyepiece is a magnifying glass, which is used to look at the real image. A strong lens is used for the eyepiece because it magnifies the image more.

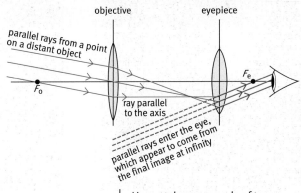

How a telescope made of two converging lenses works.

Questions

1 Why do you think Lippershey offered his invention to the Dutch military?

2 Why do telescopes and binoculars used by bird-watchers need a focusing knob?

Brighter and better

Without a telescope, you can see up to 3000 stars in the night sky. A small telescope can increase the number of visible stars to about 30 000, because it gathers more light.

The objective lens of a telescope might be 10 cm across; its **aperture** is 10 cm. On a dark night the pupil of your eye might be 5 mm across.

The amount of radiation gathered by a telescope depends on the collecting area of its objective lens. To see the faintest and most distant objects in the Universe, astronomers require telescopes with very large collecting areas.

Here is one way to gather a lot of light – build four identical telescopes. This quadruplet telescope is part of the European Southern Observatory in Chile. Each has an aperture of 8 m, so when combined they equal a single telescope with an aperture of 16 m.

Bigger and better

A telescope does not make distant stars look bigger – nearly all of them are so far away that they still appear as points of light. But it can reveal detail in objects that are either much bigger than stars (such as galaxies or clouds of glowing gas) or much closer (such as planets).

When you look at the Moon it appears quite small in the sky. Its image on the retina of your eye is quite small. Look at the Moon through a telescope, and it looks enormous. The telescope produces a greatly enlarged image on your retina.

Suppose your telescope is labelled 50 ×. This is saying that its **magnification** is 50. There is more than one way of thinking about this:

- the telescope makes the Moon seem 50 times bigger than its actual size
- the telescope makes the Moon seem to be at only 1/50th of its actual distance away
- the telescope makes the Moon's angular size look 50 times bigger.

To the naked eye, the Moon has an angular size of about half a degree (0.5°). With the telescope this is increased to 25°. The telescope has an **angular magnification** of 50.

From the ray diagram of the telescope on page 219 you can see that the angle between the rays from the eyepiece and the principal axis is larger than the angle between the rays from the object and the principal axis. This means that any **extended object**, for example the Moon, which looks small to the naked eye, will look much larger through the telescope.

A telescope does not make a distant star look bigger – it remains a point of light. However, the telescope spreads out a group of stars by magnifying the angles between them. This makes it possible to see two stars that are close together as separate objects.

Key words

- ✓ aperture
- ✓ magnification
- ✓ angular magnification
- ✓ extended object

Calculating magnification

The magnification produced by a refracting telescope depends on the lenses from which it is made.

$$\text{magnification} = \frac{\text{focal length of objective lens}}{\text{focal length of eyepiece lens}}$$

Suppose you choose two lenses with focal lengths 50 cm and 5 cm:

$$\text{magnification} = \frac{50\ \text{cm}}{5\ \text{cm}} = 10$$

Questions

3   Look at the equation for magnification above. Use it to explain why a telescope made with two identical lenses will be useless.

4   Calculate the magnification provided by a telescope made with focal lengths 1 m and 5 cm.

5   You are asked to make a telescope with as large a magnification as possible. You have a boxful of lenses. How would you choose the two most suitable lenses?

6   Show that four telescopes of diameter 8 m gather as much light as one telescope of diameter 16 m.

Find out about

✔ how radiation can be separated into different frequencies

Analysing radiation

Astronomy is not only about making images that show the size and shape of objects. By analysing the make-up of radiation received, astronomers can find out more about its source. As you will learn later in this module, radiation can provide important clues about the temperature of an object and its chemical composition. To get this information, astronomers use a **spectrometer** to measure the amount of radiation received at different frequencies. For visible light, different frequencies correspond to different colours.

Dispersion

If a beam of white light passes through a triangular block (a **prism**), the emerging ray is coloured. This is called a **spectrum**. Newton carried out a famous series of experiments with prisms. He concluded that white light is really a mixture of the colours of the spectrum. This splitting of white light into colours is called **dispersion**.

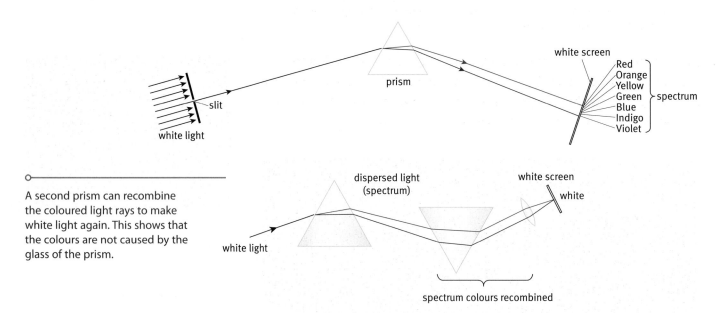

A second prism can recombine the coloured light rays to make white light again. This shows that the colours are not caused by the glass of the prism.

Newton did not know what caused the different colours, but scientists were later able to deduce that the colour of light depends on its frequency. (It therefore also depends on its wavelength, as the two are linked.) The visible band extends from red through to violet. Red has the lowest frequency (longest wavelength) and violet has the highest frequency (shortest wavelength).

Dispersion happens because light of different frequencies travels through glass (and other transparent media) at different speeds. The differences are small but they are enough to split the light up so that different colours are refracted through different angles. In glass, violet slows down more than red light, so it is refracted through a bigger angle.

Spectrometers

A spectrometer containing a prism can be attached to an optical telescope so that it produces a spectrum showing all the frequencies that are present in light.

The Pleiades is a group of bright stars. With a spectrometer, the light from each star is broken up into a spectrum.

An alternative design of spectrometer uses a **grating**, which is a set of very narrow evenly spaced parallel lines ruled on a thin sheet of glass or on a shiny surface. When light shines on the grating, different colours emerge at different angles to produce several spectra.

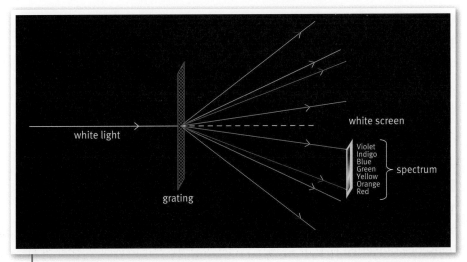

A grating produces a series of spectra.

Astronomers observing other parts of the electromagnetic spectrum also use spectrometers. Their telescopes have detectors that are only sensitive to a particular frequency. For example, radio telescopes have receivers that are tuned to specific frequencies rather like the receivers in a radio or TV.

The colours you see reflected from a CD arise because the surface has a very fine spiral track that carries the recorded information, and neighbouring sections of the track act as a grating.

Key words

- ✔ **spectrometer**
- ✔ **prism**
- ✔ **spectrum**
- ✔ **dispersion**
- ✔ **grating**

Questions

1 What are the main differences between the spectra produced by a prism and by a grating?

2 Draw a diagram to show plane waves being refracted by a prism. (Look back at the photographs of waves in ripple tanks on page 217.)

Find out about

- ✓ **the advantages of reflecting telescopes over refractors**
- ✓ **how mirrors collect light in reflecting telescopes**

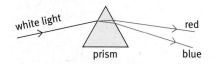

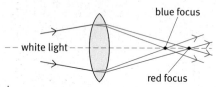

The fact that light refracts light of different colours by different amounts is useful for spectrometers but is a problem for simple lenses.

Why not lenses?

The first telescopes were **refractors** – they used lenses to refract and focus light. But most modern astronomical telescopes are **reflectors** that focus light and other electromagnetic radiation using curved mirrors.

A simple converging lens focuses different colours (frequencies) of light at slightly different points. Used as the objective lens of a telescope, it will produce an unclear image.

It is fairly easy to make small lenses that are specially designed to reduce the problem with colour. These are used in telescopes with objectives a few centimetres in diameter.

For large telescopes, lenses have some further disadvantages:

- The largest objective lens possible has a diameter of about 1 metre. Any larger and the lens would sag and change shape under its own weight, making it useless for focusing light.
- It is very difficult to ensure that the glass of a large diameter lens is uniform in composition all the way through.
- A large converging lens is quite fat in the middle. Some light is absorbed on its way through, making faint objects appear even fainter.
- Glass lenses only focus visible light. Radiation in other parts of the electromagnetic spectrum is either completely absorbed (for example, ultraviolet) or goes straight through (for example, radio).

On reflection

To focus light, a mirror must be curved. To bring parallel rays of light to a point, a mirror must be **parabolic**.

A ray diagram shows how a parabolic mirror forms a real image. It is simple to construct because each ray has to obey the law of reflection:

angle of incidence = angle of reflection

Reflecting telescopes

The first design for a reflecting telescope was proposed in 1636, not long after Galileo's refracting telescope. It used a parabolic mirror as the telescope objective.

Mirrors have several advantages over lenses, and are used in nearly all large professional telescopes.

- A mirror reflects rays of all colours in exactly the same way.
- It is possible to support a mirror several metres in diameter so that it does not sag. Its weight can be supported from the back as well as the sides.

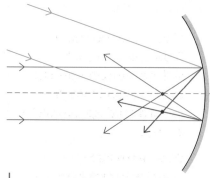

Rays parallel to the axis of the reflector are reflected to the focus. Parallel rays from another direction are focused at a different point.

- A mirror can be made very smooth so that the image is not distorted.
- By choosing suitable materials, reflectors can be made to focus most types of electromagnetic radiation.

The Arecibo radio telescope in Puerto Rico is a reflector built into a natural crater. Hanging above the dish is the detector, which is moved around to collect the reflected radio waves coming from different directions in space.

Key words
- refractors
- reflectors
- parabolic

The Arecibo radio telescope is built into a natural crater. It cannot be steered about.

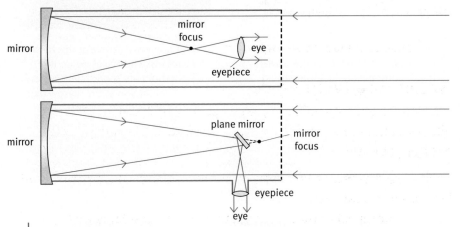

A problem with a reflecting telescope is where to place the observer. In the top diagram, the observer must be inside the telescope. In a different design, a small plane mirror close to the focus of the objective reflects light out of the telescope to an external eyepiece. Many other solutions are in common use.

Questions

1 Sketch a curved mirror. Draw two rays of light striking the mirror parallel to the axis and show how the mirror brings the rays to a focus.

2 Write down three advantages of using a reflector telescope rather than a refractor telescope.

Find out about

- ✔ **what happens to waves when they go through an aperture**
- ✔ **why radio telescopes are in general much bigger than optical ones**

Clear image

Most modern professional telescopes are very big. You have already considered one reason for this: they need to collect a lot of radiation in order to detect faint objects. But there is another reason: it is to make the image clearer so that more detail can be seen.

If an image is blurred, you cannot resolve (distinguish) much detail. To get a clear image you need a telescope with good **resolving power**.

Radio telescopes have to be much bigger than optical ones to get the same resolving power. To understand why, you need to study what happens to waves when they go through an aperture.

Waves through an aperture

When waves hit a barrier, they bend a little at the edge and travel into the shadow region behind the barrier. This effect is called **diffraction**. The longer the wavelength of the wave, the more it diffracts.

At a gap between two barriers, waves bend a little at both edges. If the width of the gap is similar to their wavelength, the waves beyond the gap are almost perfect semicircles. If the gap is really tiny, much less than the wavelength of the waves, the waves do not go through at all.

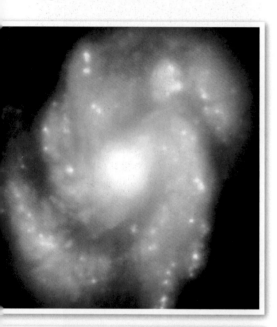

Two views of a region of the night sky, taken with telescopes with different resolving powers.

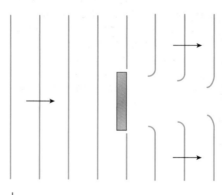

Diffraction occurs when waves meet an obstacle. The edges of the waves bend round the obstacle, into the shadow region behind.

Plane waves arrive at this harbour mouth. The waves inside the harbour are semicircular, because of diffraction at the narrow gap between the two piers.

Waves spread out as they pass through the aperture.

A narrower aperture has more effect.

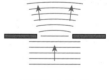

A smaller wavelength gives less diffraction.

The effect is greatest when the aperture is similar to the wavelength of the waves.

Diffraction and resolving power

When a beam of radiation arrives from a very distant object ('at infinity'), it consists of plane waves all travelling in one direction. But when the beam goes through the aperture of a telescope (or the pupil of your eye), diffraction occurs and the waves spread out to give a blurred image.

Even a very small amount of diffraction causes slight blurring. So to reduce diffraction effects, astronomers design telescopes with apertures much larger than the wavelength of the radiation they want to gather.

For optical astronomers, diffraction is not too big a problem because the wavelength of light is so small. More blurring is usually caused by 'twinkling' as light passes through the atmosphere.

Diffraction is much more of a problem for radio astronomers. The wavelengths they use are normally a few centimetres. Even a radio telescope a mile across does not have very good resolving power.

To get a big aperture, radio telescopes often have an array of several small dishes linked together. This arrangement can have the same resolving power as a single telescope with an aperture equal to the distance across the array.

The Very Large Array in New Mexico, USA, is the world's largest radio telescope. It has 27 dishes that can be moved along the 21-km-long arms of a Y-shaped track.

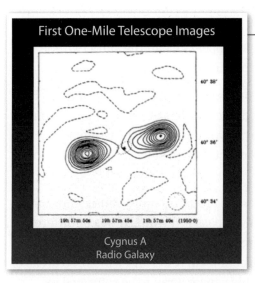

First One-Mile Telescope Images

Cygnus A
Radio Galaxy

This image of the galaxy Cygnus A was made in the 1960s with a one-mile radio telescope (its dishes were spread out over a distance of one mile). It is shown as a contour map with lines joining points of equal brightness. Compare it with the image of the same galaxy on page 211.

Key words
- ✓ **resolving power**
- ✓ **diffraction**

Questions

1   Draw diagrams to show what happens to a plane wave as it passes through a gap in a barrier that is:
    a   bigger than its wavelength
    b   about the same size as its wavelength.

2   Look at the two radio images of the galaxy Cygnus A on this page and on page 211. In which image are the details better resolved?

3   An astronomer finds that she can scarcely resolve two images. Which of these will improve the situation:
    a   using a telescope with a smaller aperture?
    b   observing radiation of a shorter wavelength?
    Explain your answer.

Some telescopes are built on high mountains. Others are carried on spacecraft in orbit around the Earth. This is partly because some types of electromagnetic radiation are absorbed by the atmosphere.

- The atmosphere **transmits** (allows through) visible light, microwaves, radio waves, and some infrared.
- Other radiation, including X-rays, gamma rays, and much infrared, is **absorbed**.

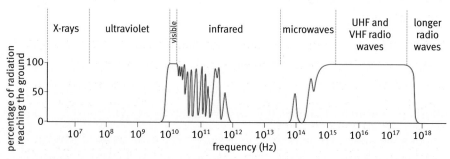

Graph showing the percentage of radiation reaching the Earth's surface across the electromagnetic spectrum.

The Hubble Space Telescope orbits above the Earth, avoiding the effects of atmospheric absorption and refraction of light. As well as gathering visible light, Hubble also gathers ultraviolet and infrared radiation.

Twinkle, twinkle

If you look up at the stars in the night sky, you are looking up through the atmosphere. Even when the sky appears clear, the stars twinkle. This shows an important feature of the atmosphere.

The stars, of course, shine more or less steadily. Their light may travel across space, uninterrupted, for millions of years. It is only on the last few seconds of its journey to your eyes that things go wrong. The twinkling, or scintillation, is caused when starlight passes through the atmosphere.

The atmosphere is not uniform. Some areas are more dense, and some areas are less dense. As a ray of light passes through areas of different densities, it is refracted and changes direction. The atmosphere is in constant motion (because of convection currents and winds), so areas of different density move around. This causes a ray to be refracted in different directions, and is the cause of scintillation.

Some astronomical telescopes record images electronically. Computer software allows them to reduce or remove completely the effects of scintillation from the images they produce.

Dark skies, please

A telescope 'sees' stars against a black background. However, many astronomers find that the sky they are looking at is brightened by **light pollution**. Much of this is light that shines upwards from street lamps and domestic lighting. **Scattered** by the atmosphere, the light enters any nearby telescope.

Another consequence is that it has become difficult to see the stars at night from urban locations. People today are less aware of the changing nature of the night sky, something that was common knowledge for our ancestors.

In 2006, the city authorities in Rome decided that enough was enough. To reduce light pollution and to save energy, it was decided to switch off many of its 170 000 street lights, thereby cutting its lighting bill by 40%. Illumination of its ancient monuments has been dimmed, as well as lights in shop and hotel windows.

The Campaign for Dark Skies, supported by amateur and professional astronomers, campaigns to reduce light pollution. But it is not only light radiation that affects astronomers. Electrical equipment can produce weak radio waves, particularly when being switched on and off. And radio waves used for broadcasting and mobile phones can interfere with the work of radio telescopes, so certain ranges of frequencies must be left clear for astronomical observations.

These two maps show the amount of light pollution across the United Kingdom in 1993 and 2000. The red areas indicate where most light is emitted, the dark-blue areas where the least is emitted. You can see that the red areas have grown over the seven years between surveys. But the biggest change is in the countryside. Here the light pollution, although at lower levels than in towns and cities, has increased across large areas of England.

Questions

1 Some astronomers want to study radiation with a frequency around 10^9 Hz. Use information from the graph to explain whether they could do this with a ground-based telescope.

2 Suggest how information from a space telescope could be sent to astronomers on the ground.

3 Do you think it is important to reduce light pollution? Give reasons for your point of view.

Astronomers working together at ESO

There are parts of the sky that are not visible from the northern hemisphere. So 14 European countries have joined together to do astronomical research in an organisation known as the European Southern Observatory (ESO). Together they operate observatories at three sites in the Atacama Desert, high up in the Andes mountains of Chile. Two of ESO's sites are on mountain tops at heights of about 2500 m above sea level. The third observatory, which ESO is building together with other international partners, is the ALMA array of radio telescopes on a 5000-m-high plateau. These remote locations are chosen because the atmosphere is both clear and dry there. This reduces the effects of absorption and refraction of radiation, and of light pollution.

Remote control

Each year, astronomers make observations for about 500 projects using the ESO telescopes. Half of them do this from their home base in Europe. ESO does this by making great use of the power of computers and the Internet. The telescopes in Chile are remote: it takes two days of travel to reach them, and two days to get home again. So many of the astronomers who need observations to be made send in their requests, and these are programmed into the ESO control system. Local operators ensure that the observations are made, and the results are sent back to Europe.

Aerial view of ESO's Paranal Observatory, Chile. The summit of Cerro Paranal is at an altitude of 2600 m.

One benefit of this is to avoid the 'weather lottery'. ESO telescope sites were chosen for their excellent weather but some observations are so sensitive that they require the clearest of skies that these places have to offer. Imagine travelling of Chile for three nights' observing, only to find that the skies were not quite clear enough! ESO's computer control ensures that your observations are postponed to a later date, while another astronomer's observations are brought forward to take advantage of the available time.

ESO's flagship telescope is the Very Large Telescope (VLT) on the 2600-m-high mountaintop of Cerro Paranal in Chile. It is not just one, but an array of four giant telescopes, each with a main mirror 8.2 m in diameter. The next step beyond the VLT is to build a European Extremely Large Telescope (E-ELT) with a main mirror 42 m in diameter. The E-ELT will be "the world's biggest eye on the sky" — the largest telescope in the world to observe visible light and infrared light, and ESO is currently drawing up detailed design plans.

The ESO Very Large Telescope (VLT) on Cerro Paranal in Chile.

The Paranal Residencia accommodates visiting astronomers. The central area is naturally lit through a dome in the ceiling. At the centre of the dome, an umbrella-shaped blackout curtain automatically opens at sunset, to avoid artificial light escaping and interfering with astronomical observations. The curtain closes automatically at sunrise.

Control room for the ESO Very Large Telescope.

ESO people

Douglas Pierce-Price is a British astronomer based at the ESO headquarters in Germany.

How many people work for ESO?

In 2010 ESO employs about 740 staff members around the world (primarily at our headquarters in Germany and in Chile). Included in this are about 180 local staff recruited in Chile, thus providing work with high skill levels for local people.

You don't have to be an astronomer to work at ESO: we also employ engineers, including software engineers, as well as technical and administrative staff. In addition, there are currently opportunities annually for about 40 students and 40 research fellows who are attached to the organisation. These opportunities are open either to students enrolled in a Ph.D. programme or Fellows who have achieved their Ph.D. in astronomy, physics, or a related discipline.

What is it like to live and work in a desert on top of a mountain?

When working at ESO's Very Large Telescope (VLT) on Cerro Paranal, the observatory staff live and work in the Residencia, a futuristic building built partly underground and with a 35-metre-wide glass dome in the roof. It is part of the VLT's 'base camp' facility, situated a short distance below the summit of Cerro Paranal.

Astronomical conditions at Paranal are excellent, but they come at a price. It's a forbidding desert environment; virtually nothing can grow outside. The humidity can be as low as 10%, there are intense ultraviolet rays from the Sun, and the high altitude can

Douglas Pierce-Price

leave people short of breath. The nearest town is two hours away, so there is a small paramedic clinic at the base camp.

Living in this extremely isolated place feels like visiting another planet. Within the Residencia, a small garden and a swimming pool are designed to increase the humidity inside. The building provides visitors and staff with some relief from the harsh conditions outside: there are about 100 rooms for astronomers and other staff, as well as offices, a library, cinema, gymnasium, and cafeteria.

Up all night

Monika Petr-Gotzens is a German astronomer working at the European Southern Observatory. She is studying how stars form in dense clusters. She is particularly interested in the formation of binary stars. These are pairs of stars that orbit one another. Almost half of all stars are in binary pairs.

Monika uses both radio telescopes and optical (light) telescopes in her work.

What is it like to work with a telescope at the top of a mountain?

The Observatory sites are in such isolated places where, during new-moon, the nights are the darkest nights I have ever seen, and the work at the mountain top is accompanied by a natural, amazing silence that one can only experience at these very remote sites. When standing outside the telescope domes you hear only a low drone from the telescope while it changes its pointing position. It is here, on the telescope platform, that one gets the feeling of being part of a very special mission.

The way that we observe with telescopes today is quite different from the image that many people have of an astronomer standing in the cold with their eye to a telescope, looking towards the sky. The control over the telescope and instrument is 100% computer based, and carried out from a control room. The control room is about 100 m away from the telescope, and you don't even see the telescope during your observation, unless you actively walk into the dome. So it isn't the freezing cold that keeps you awake during the observing night, but the smell of coffee and cookies.

What part do computers play in your work?

Modern astronomy without computer control is unthinkable. It sharpens our view of the Universe by directing the telescope very accurately. Computers also process the data gathered by the telescope to give much higher quality images and measurements.

Astronomers today usually work in collaborative teams. Why is this?

International collaborations are very important. Nowadays, it is not a single clever observation that solves one of the grand questions of the Universe. They are answered through major efforts. For example, very deep surveys of stars or galaxy fields across large areas of the sky. It is often impossible for

Monika Petr-Gotzens

individuals to deal with the huge amount of data accumulated from such surveys.

The ESO telescopes are sited high on mountains. Why is this?

The factors that influence the choice of site for an observatory depend on the kind of observatory: optical, infrared, millimetre, or radio waves. For optical observatories, a low-turbulent atmosphere (i.e. very good seeing), a dark sky without light pollution, and a high number of clear sky nights are important factors. It's easier for radio telescopes – they can see through clouds to some extent.

The size of the telescope is also important. Independent of the working wavelength, the rule applies that the larger the telescope, the less windy a site must be, to avoid image blur from vibrations caused by wind shake.

On top of that, sites must also have a practical logistical supply. It must be reasonably easy for astronomers to travel to the observatory, and they need accommodation, food, and drink. Natural springs, for example, for the water supply, are an advantage, although not an absolute requirement.

This spiral galaxy looks rather like the Milky Way seen from the outside. It is 25 million light-years away.

Astronomical distances

Astronomy deals with gigantic objects lying at mind bogglingly huge distances. The pictures on these two pages give just a few examples.

Solar System

Neptune is the furthest major planet from the Sun. The radius of its orbit is about 4500 million km. Beyond Neptune there are dwarf planets and other small objects, such as comets, that orbit the Sun so are part of the Solar System. The furthest of these could be as far as one light-year from the Sun.

One light-year (ly) is the distance electromagnetic radiation travels through space in one year. $1 \text{ ly} = 9.5 \times 10^{12} \text{ km}$

Milky Way galaxy

The Sun is one of about 100 thousand million stars that make up the Milky Way galaxy. This disc-shaped collection of stars is about 100 000 light-years across, and the Sun lies about two-thirds of the way out from the centre.

We can't see the Milky Way from the outside, but we can see other galaxies. Like the Milky Way, some are disc shaped with their brightest stars tracing out a spiral shape.

There is more to the Solar System than eight planets and the Sun. This image is created using data about minor planets, asteroids and comets. The four outer planets are blue. Asteroids associated with Jupiter's orbit are pink. The green objects are small objects such as comets and minor planets. The pronounced gap at the bottom is due to lack of data because the sky is obscured by the band of the Milky Way.

Other galaxies

Some galaxies are spirals like the Milky Way, but many are simple oval-shaped collections of stars. Some galaxies have strange shapes that might have been produced when two galaxies collide and merge.

This pair of colliding spiral galaxies is about 400 million light-years away.

This elliptical galaxy is about 300 million light-years away.

The Universe

Since the early 20th century, telescopes have revealed millions of galaxies scattered through space at distances up to 13 *thousand* million light-years. The light we are receiving from them now must have been emitted 13 thousand million years ago – not long after the Big Bang that began the Universe.

How do we know?

In order to understand and explain the Universe and the objects we see in it, we need to know how far away they are. Once an object's distance is known, then astronomers can work out other things, such as how big it is and how much energy it is giving out.

Distance measurement is one of astronomy's major challenges. You can't simply pace out the distance or use a tape measure. Nor (with the exception of the Moon and some nearby planets) is it possible to send a radar pulse and time how long it takes before you detect its echo. So astronomers have had to rely on observations of radiation received on Earth, coupled with ingenuity. In this topic you will learn more about some of the methods they have devised.

The Hubble Space Telescope has given us a striking view of Universe that contains many billions of galaxies.

Parallax angles

The stars are far off. How can we measure their distances? One way is to use the idea of parallax.

Imagine looking across a city park in which there are a number of trees scattered about. You take a photograph. Now take two steps to the right and take another photograph. Your photos will look very similar, but the *relative* positions of the trees will have changed slightly. Perhaps one tree that was hidden behind another has now come into view.

These two photos illustrate the effect of parallax. They show the same view, but the photographer moved sideways before taking the second one. The closest object, the person on the bench, has moved furthest across the image.

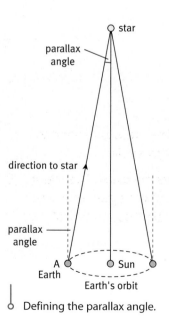

Defining the parallax angle.

Now superimpose the photos one on top of the other and you will see that the closer trees will have shifted their positions in the picture more than those that are more distant. You have observed an effect of **parallax**.

Astronomers can see the same effect. As the Earth travels along its orbit round the Sun, some stars seem to shift their positions slightly against a background of fixed stars. This shifting of position against a fixed background is what astronomers call parallax, and it can be used to work out the distance of the star in question.

The diagram on the left shows how astronomers define the **parallax angle** of a star. They compare the direction of the star at an interval of six months. The parallax angle is *half* the angle moved by the star in this time.

From the diagram, you should be able to see that, the closer the star, the greater is its parallax angle.

The scale of things

In the Middle Ages, astronomers imagined that all of the stars were equally distant from the Earth. It was as if they were fixed in a giant crystal sphere, or perhaps pinholes in a black dome, letting through heaven's light. They could only believe this because the patterns of the stars do not change through the year – there is no obvious parallax effect.

They were wrong. But it is not surprising that they were wrong, because parallax angles are very small. The radius of the Earth's orbit is about 8 light-minutes, but the nearest star is about 4 light-years away – that's over 250 000 times as far.

Parallax angles are usually measured in fractions of a second of arc. There are:

- 360° in a full circle
- 60′ (minutes) of arc in 1°
- 60″ (seconds) of arc in 1′

So a second of arc is $\frac{1}{3600}$ of a degree.

Astronomers use a unit of distance based on this: the **parsec** (pc).

- An object whose parallax angle is 1 second of arc is at a distance of 1 parsec.

$$\text{Distance (parsec)} = \frac{1}{\text{parallax angle (sec)}}$$

Because a smaller angle means a bigger distance:

- An object whose parallax angle is 2 seconds of arc is at a distance of 0.5 parsec.

A parsec is about 3.1×10^{13} km. This is of a similar magnitude to a light-year, which is 9.5×10^{12} km. Typically, the distance between neighbouring stars in our galaxy is a few parsecs.

The European Hipparcos satellite measured the parallax angles of over 100 000 stars out to a distance of about a thousand parsecs.

Key words

- ✓ **parallax**
- ✓ **parallax angle**
- ✓ **parsec**

Questions

1 Draw a diagram to show that a star with a large parallax angle is closer than one with a small parallax angle.

2 How many light-years are there in a parsec?

3 What is the parallax angle of a star at a distance of 1000 pc?

4 If a star has a parallax angle of 0.25 seconds of arc, how far away is it (in parsecs)?

5 Suggest reasons why a satellite was needed to measure very small parallax angles.

6 Suggest why astronomers calculate distances in parsecs, but newspapers and magazines use the light-year when writing about astronomical distances.

Brightness and distance

Measurements of parallax angles allow astronomers to measure the distance to a star. This only works for relatively nearby stars. But there are other methods of finding how far away a star is.

In the late seventeenth century, scientists were anxious to know just how big the Universe was. The Dutch physicist Christiaan Huygens devised a technique for measuring the distance of a star from Earth. He realised that, the more distant a star, the fainter its light would be. This is because the light from a star spreads outwards, and so the more distant the observer, the smaller the amount of light that they receive. So measuring the **observed brightness** of a star would give an indication of its distance.

Here is how Huygens set about putting his idea into practice:
- At night, he studied a star called Sirius, the brightest star in the sky.
- The next day he placed a screen between himself and the bright disc of the Sun. He made a succession of smaller and smaller holes in the screen until he felt that the speck of light he saw was the same brightness as Sirius.
- Then he calculated the fraction of the Sun's disc that was visible to him. It seemed that roughly 1 / 30 000 of the Sun's brightness equalled the brightness of Sirius. His calculation showed that Sirius was 27 664 times as distant as the Sun.

Huygens understood that his method had some problems. Here are three of them:
- First, there was subjectivity in his measurements. He had to judge when his two observations through the screen were the same.
- Second, his method assumed that Sirius and the Sun are identical stars, radiating energy at the same rate.
- Third, he had to assume that no light was absorbed between Sirius and his screen.

Astronomers now know that Sirius is about 500 000 times as distant as the Sun. But at the time, Huygens' measurement was a breakthrough because it used the idea that the Sun would look like other stars if seen from far enough away. Also, Huygens was the first to show that stars lay at such vast distances.

Light spreads out, so the more distant a source is the less bright it appears.

Luminosity

Stars are not all the same. They do not all give out the same amount of light – they have different luminosities.

The **luminosity** of a star is its power output. It is the energy it gives out each second by radiating light and other types of electromagnetic radiation. The Sun's luminosity is about 4×10^{26} W – compare this with the 'luminosity' of a typical electric lamp.

A star's luminosity depends on two factors:

- its temperature – a hotter star radiates more energy per second from each square metre of its surface
- its size – a bigger star has a greater surface radiating energy.

The observed brightness recorded by an astronomer depends on a star's luminosity as well as its distance. Also, any dust or gas between Earth and the star may absorb some of its light.

If astronomers are confident that two stars have the same luminosity, they can use their observed brightness to compare the stars' distances. But rather than relying on subjective judgement like Huygens did, they use sensitive instruments to measure brightness.

Key words

✓ **observed brightness**
✓ **luminosity**

Questions

7 Suggest at least one other problem with Huygens' method.

8 Explain how two stars having the same observed brightness may have different luminosity.

9 If some starlight is absorbed by dust, explain whether this would make a star appear closer or further away than it really is.

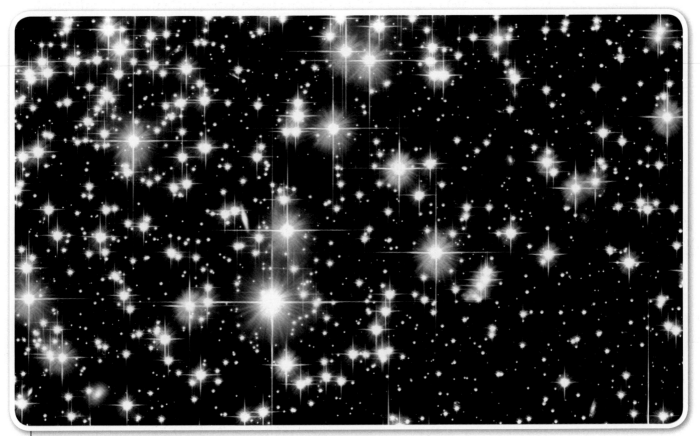

These stars are all roughly 8200 light-years from Earth. They have a range of luminosities, giving a range of observed brightness.

Find out about

- ✔ why some stars vary regularly in brightness
- ✔ how Cepheid variable stars can be used to measure astronomical distances

Cepheid variable stars

The Sun and many other stars emit light at a steady rate. But some stars are **variable stars**, meaning that their luminosity varies. Variations in luminosity can provide clues about what's happening in the stars – and can sometimes have other uses too.

In 1784 a young English astronomer, John Goodricke, discovered a new type of variable star. He noticed that a star called δ Cephei (δ = delta) went from dim to bright and back again with a time **period** of about a week, and that this variation was very regular. The graph below shows some modern measurements of the brightness of this star.

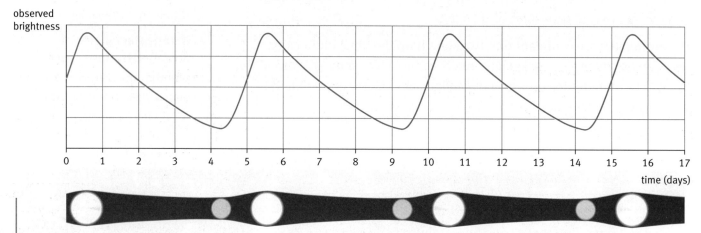

The observed brightness of δ Cephei varies regularly. The variation in its luminosity is caused by its expansion and contraction.

Many stars have been found that vary in this way, and they have been named Cepheid variables, or simply **Cepheids**. It is now thought that a star like this is expanding and contracting so that its temperature and luminosity vary. Its diameter may vary by as much as 30%.

Cepheids and distance measurement

In the early years of the twentieth century, an American astronomer called Henrietta Leavitt made a very important discovery. She looked at Cepheids in a small galaxy close to the Milky Way. She noticed that the brightest Cepheids varied with the longest periods, and drew a graph to represent this.

Because the stars she was studying were all roughly the same distance, Leavitt realised that the stars that appeared brightest were also the ones with the greatest luminosity – they were not brighter simply because they were closer.

Henrietta Leavitt, whose work opened up a new method of measuring the Universe.

Measuring distances

Henrietta Leavitt had discovered a method of determining the distance to star clusters in the Milky Way, and to other galaxies. There are two parts to the method.

Part one:
- find some nearby Cepheids whose distances have been measured using other methods
- measure their brightness and work out their luminosities
- plot a graph of luminosity against period.

Part two:
- look for a Cepheid in a star cluster or galaxy of interest
- measure its observed brightness and period of variation
- from the period, read its luminosity off the graph
- use the luminosity and observed brightness to work out the distance.

The Cepheid method has been used to measure distance to galaxies up to a few **megaparsecs** (Mpc) away. (1 Mpc = 1 million parsecs)

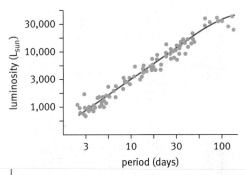

The luminosity of a Cepheid variable star is related to its period.

Questions

1 From the graph on the previous page, deduce the period of variation of δ Cephei.

2 Why did Henrietta Leavitt assume that the stars she was studying were all at roughly the same distance from Earth?

3 Suggest how distances to nearby Cepheids might have been measured.

4 Suggest why Cepheids cannot be used to measure distances beyond a few Mpc.

Key words
- ✓ variable stars
- ✓ period
- ✓ Cepheids
- ✓ megaparsecs

The shape of the Milky Way

On a clear dark night, you can see a faint milky band of light stretching across the sky. This is the Milky Way. With a telescope, you can see that its light comes from vast numbers of faint stars surrounding us in a disc-shaped arrangement.

In 1785, William Herschel attempted to determine the shape of the **galaxy**. Looking through his telescope, he counted all the stars he could see in a particular direction. Then he moved his telescope round a little and counted again. Once he had done a complete circle, he could draw out a map of a slice through the Milky Way.

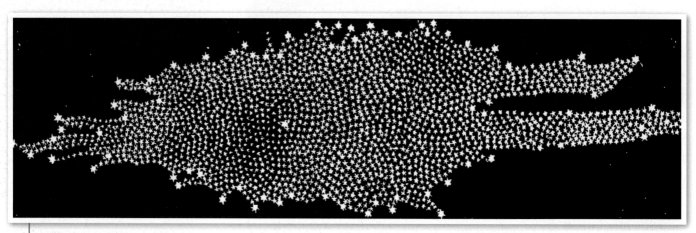

William Herschel's map of a slice through the Milky Way. The Sun is shown near the centre. The more stars seen in a particular direction, the greater the distance to the edge in that direction.

Herschel knew that he was making the following assumptions:

- that his telescope could detect all the stars in the direction he was looking
- that he could see to the far end of the galaxy.

Herschel himself discovered his first assumption to be incorrect when he built a bigger telescope. Astronomers now know that Herschel's second assumption was also incorrect. Dust in the galaxy makes it difficult to see stars in the packed centre of the galaxy.

Shapley and the nebulae

By the early 20th century, astronomers were starting to use really big telescopes, especially in America. Harlow Shapley worked in California, investigating faint patches of light called nebulae (**nebula** means cloud; nebulae is the plural). Some nebulae are irregular 'blobs' of light, some are roughly circular, while others are spirals. A good telescope reveals that some nebulae are gas clouds but others are clusters of stars.

Thanks to Henrietta Leavitt, Shapley had a new way to measure distances to stars in nebulae. He found that the roughly spherical star

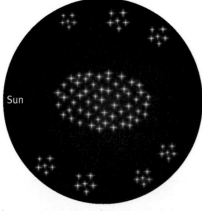

Harlow Shapley's idea of the Milky Way galaxy.

Sun

clusters (now called **globular clusters**) had distances up to about 100 000 light-years, and that they seemed to lie in a sphere around the Milky Way.

Spiral nebulae – the great debate

Shapley and some other astronomers thought the spiral nebulae were in the Milky Way. Others, including Heber D Curtis (also American) suggested they were 'island universes' a very long way outside the Milky Way.

In 1920, Shapley and Curtis held a public 'great debate'. On the night, Curtis came off better, partly because he was the better speaker.

Then in 1923, American astronomer Edwin Hubble was using a new telescope to study the Andromeda nebula. He spotted a faint Cepheid and found it was almost one *million* light-years away. This was enough to convince astronomers that they were looking at a separate, distant galaxy. Other spiral nebulae were then also found to be galaxies.

This artist's impression of the Milky Way is based on recent optical and infrared measurements.

This spiral nebula is in the part of the sky belonging to the constellation of Andromeda.

A recent view

Modern optical and infrared telescopes have been used to map the layout of stars in the Milky Way. They show that, like the Andromeda nebula, it is a spiral galaxy.

Questions

1 Suggest reasons why Shapley's evidence seemed stronger than Curtis's on the night of the great debate.

2 What new observations showed that there were objects outside the Milky Way?

Find out about

- ✔ some of the types of object studied by astronomers
- ✔ typical sizes and distances of some objects studied by astronomers
- ✔ why supernovae can be used to measure distance

More than just stars

Astronomers study many different types of object, not just stars and planets. Measurements of their distances and sizes help us to build our picture of the Universe.

Some of these objects are found in the Solar System. In astronomical terms, they are relatively nearby. Other objects of interest are found within the Milky Way galaxy. And others are seen at vast distances far beyond the Milky Way. The images on this page show just some examples of objects discovered and studied by astronomers.

Comets travel around the Sun in very long thin orbits. This is Halley's **comet**, which passes close to Earth every 76 years. At its furthest point, it reaches about 5 thousand million km from the Sun.

In 2010 a Japanese spacecraft returned to Earth after visiting the **asteroid** Itokawa, one of thousands of asteroids orbiting the Sun between Mars and Jupiter.

Key words

- ✔ comet
- ✔ asteroid
- ✔ planetary nebula
- ✔ supernova
- ✔ supernova remnant

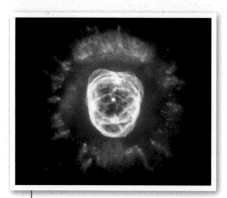

The Eskimo (or Clownface) nebula, about 1000 parsecs (3000 light-years) from Earth, is a **planetary nebula**. This name is misleading because they are nothing to do with planets. They are formed when a dying Sun-like star throws off its outer layers of hot gas. You can often see the remains of the star at the centre of the nebula.

The Crab nebula is about 2000 parsecs from Earth. It is the remains of a dying star that exploded in 1054 ad. The explosion, called a **supernova**, was so powerful that it could be seen in daylight from Earth. At the centre of the nebula is a pulsar (see page 210), all that remains of the original star. The nebula is an example of a **supernova remnant**.

This **quasar** is about 5 hundred million parsecs from Earth. Quasars are some of the most distant objects that can be seen from Earth. They are thought to be galaxies containing gigantic black holes that draw in material from their surroundings, heating it so that it emits vast amounts of radiation.

What can we learn from supernovae?

Supernova explosions are rare. They happen about once per century in a typical spiral galaxy. The most recent one seen in the Milky Way was in 1604 ad, and in 1987 one was observed in a small galaxy close to the Milky Way, about 168 000 light-years from Earth.

Astronomers study supernovae and their remnants in order to learn more about how stars come to an end. But there is another reason too.

Supernova explosions are monitored by measuring how their light output varies with time. One particular type of supernova always produces the same shape graph, and always seems to reach roughly the same peak luminosity. Astronomers can use these supernovae to work out the distances to galaxies where they are observed, in the same way that they use Cepheids to measure distance.

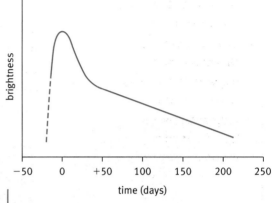

Graph showing how light output from a supernova explosion varies with time.

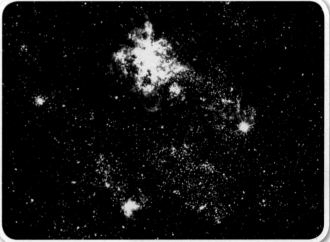

This supernova (shown in the bottom image on the right) was seen in February 1987, two days after it exploded. (The upper image was taken in 1969.) It is the closest supernova that has been seen for over 300 years.

Questions

1 For each object shown on these pages, say whether it is found within the Solar System, within the Milky Way galaxy, or far beyond the Milky Way. Use information from pages 234 and 235 to help you decide.

2 Write a plan for using supernovae to measure distance. Base your plan on the one for Cepheids on pages 240 and 245.

3 Suggest one reason for using supernovae, rather than Cepheids, to measure distances. Suggest at least one problem with using supernovae.

Find out about

- ✓ the discovery of the recession of galaxies
- ✓ evidence for the expansion of the Universe

Edwin Hubble using the 48-inch telescope at the Mount Palomar observatory.

Moving galaxies

Edwin Hubble was fortunate to be working at a time (the 1920s) when the significance of Cepheid variables had been realised. They could be used as a 'measuring stick' to find the distance to other galaxies. At the same time, he was able to use some of the largest telescopes of his day, reflectors with diameters up to 5 metres.

Hubble conducted a survey of galaxies, objects that had not previously been seen, let alone understood, until these powerful instruments became available. In his book *The Realm of the Nebulae* (1936), he described what it was like to see individual stars in other galaxies:

> The observer looks out through the swarm of stars that surrounds him, past the borders and across empty space, to find another stellar system . . . The brightest objects in the nebula can be seen individually, and among them the observer recognises various types that are well known in his own stellar system. The apparent faintness of these familiar objects indicates the distance of the nebula – a distance so great that light requires seven hundred thousand years to make the journey.

Redshift

Hubble used Henrietta Leavitt's discovery to determine the distance of many galaxies. At the same time, he made a dramatic discovery of his own. This was that the galaxies all appeared to be receding (moving away) from us. He deduced this by looking at the spectra of stars in the galaxies. The light was shifted towards the red end of the spectrum, a so-called redshift.

It turned out that, the more distant the galaxy, the greater its **speed of recession** – another linear relationship. Hubble's graph shows that, although his data points are scattered about, the general trend is clear.

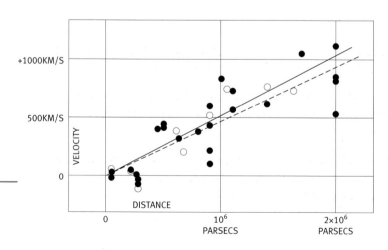

Edwin Hubble's graph, relating the speed of recession of a galaxy to its distance.

The Hubble constant

Hubble's finding can be written in the form of an equation:

speed of recession = Hubble constant × distance

The quantity called the **Hubble constant** shows how speed of recession is related to distance. His first value (from the graph) was about 500 km/s per megaparsec. In other words, a galaxy at a distance of 1 Mpc would be moving at a speed of 500 km/s. A galaxy at twice this distance would have twice this speed.

Other astronomers also began measuring the Hubble constant, using many more distant galaxies. Hubble's first value was clearly too high. For decades, the measurement uncertainties remained high and so there were disputes about the correct value. By 2010, the accepted value of the Hubble constant was 70.6 ± 3.1 km/s per Mpc. This value is based on a large number of measurements, including using many Cepheids and supernovae.

Back to the big bang

Astronomers had discovered that clusters of galaxies are all moving apart from each other. The further they are away, the faster they are moving. It doesn't matter where you are in the Universe, everything appears to moving away – space itself seems to be expanding. Edwin Hubble's discovery:

- the Universe itself may be expanding, and may have been much smaller in the past
- the Universe may have started by exploding outwards from a single point – the big bang.

This is now the widely accepted model of how the Universe began. Scientists use the model to work out how long ago the expansion started. They calculate that the Universe is about 14 thousand million years old.

Questions

1 Calculate the speed of recession of a galaxy that is at a distance of 100 Mpc, if the Hubble constant is 70 km/s per Mpc.

2 Using the same value of the Hubble constant given in question 1, calculate the distance of a galaxy whose speed of recession is 2000 km/s.

3 A galaxy lies at a distance of 40 Mpc from Earth. Measurements show its speed of recession is 3000 km/s. What value does this suggest for the Hubble constant?

Key words
- ✓ speed of recession
- ✓ Hubble constant

Topic 4: Stars

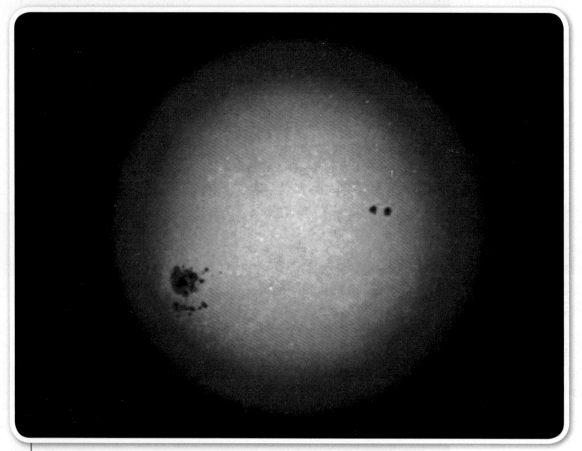

The surface of the Sun is quite dramatic when seen from close up. These are sunspots, cooler areas of the surface. They are still very hot, but perhaps 1000 °C cooler than the average surface temperature of about 5500°C.

Mystery of the Sun

The Sun is our star. By understanding the Sun better, astronomers hope to be able to make more sense of the variety of stars they see in the night sky.

The Sun is much closer than any other star, so it is the easiest star to gather detailed scientific data on. Scientists study its surface and analyse the radiation coming from it. They turn their telescopes on it, and send spacecraft to make measurements from close up.

An energy source

What makes the Sun work? How does it keep pouring out energy, day after day, year after year, millennium after millennium – for billions of years? Why does it burn so steadily? These are questions that have puzzled scientists for centuries.

Some suggestions were that it was powered by volcanoes, that it was burning coal, that it used the energy of comets that fell into it, or simply the energy of the Sun itself as it collapsed inwards under the pull of its own gravity. In this topic you will find out more about the energy source that powers the Sun and other stars.

Composition

In the 1830s, the French philosopher Auguste Comte suggested there were certain things that we could never know. As an example, he gave the chemical composition of the stars. But by 1860 physicists had interpreted the spectrum of starlight and identified the elements present. In this topic you will learn how this works.

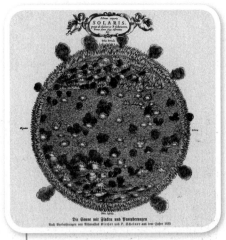

Seventeenth century view – the Sun as an enormous lump of coal.

The dark lines in this spectrum show which chemical elements are present in the Sun.

Weather from the Sun

The Sun seems to shine steadily, but actually it is a place of violent upheaval. Detailed observations reveal giant bubbles of gas bursting out of the Sun and flying out into space. These are now known as coronal mass ejections (CMEs) and weigh as much as 100 billion tonnes.

CMEs consist of electrically charged particles. When one reaches Earth it can produce dramatic aurora effects in the night sky, but it can also damage electrical communication and power supply systems.

A CME can reach Earth in about four days, travelling at 1.5 million km/h. Light takes just 500 s to make the same journey, so space 'weather forecasts' can give advance warning of a CME's arrival.

In this topic you will learn about some of the processes that take place as stars form, as they shine, and as they end their lives.

A coronal mass ejection (CME) photographed from the orbiting SOHO solar observatory using an opaque disc to block the Sun's bright disc.

Find out about

✔ **how colour is related to temperature**
✔ **how the spectrum of radiation from a hot object depends on its temperature**

Colour and temperature

A star produces a continuous range of frequencies across the **electromagnetic spectrum**. Most of its radiation is in the infrared, visible and ultraviolet regions of the spectrum, but stars also produce radio waves and X-rays. If you look out at the stars at night, you may get a hint that they shine with different colours – some reddish, some yellow (like the Sun), others brilliantly white. It is more obvious if you look through binoculars or a telescope.

Colour is linked to temperature. Imagine heating a lump of metal in a flame. At first, it glows dull red. As it gets hotter, it glows orange, then yellow, then bluish white.

You might notice that these colours appear in the order of the spectrum of visible light. Red is the cool end of the spectrum, violet the hot end. For centuries, the pottery industry has measured the temperature inside a kiln by looking at the colour of the light coming from inside.

So the colour of a star gives a clue to its surface temperature.

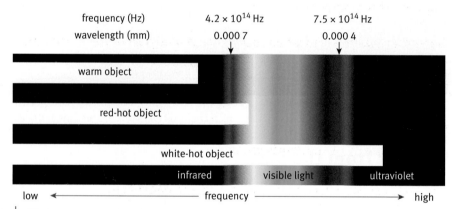

All objects emit some electromagnetic radiation. As the temperature rises, the amount of radiation at high frequencies increases.

Analysing starlight

At one time, astronomers judged the colours of stars and classified them accordingly. However, it is better to analyse stars' light using an instrument called a spectrometer. A spectrometer can be attached to a telescope so that it produces a spectrum, showing all of the frequencies that are present. The photographs show how a spectrometer turns the light from each star into a spectrum.

The Pleiades (top) is a group of bright stars about 500 light-years away. With a spectrometer, the light from each star is broken up into a spectrum (bottom image), revealing slight differences in their colour.

Comparing stars

Better still is to turn the spectrum of a star into a graph. The graph below shows the intensity (energy radiated per unit area of a star's surface) for each frequency in the spectrum.

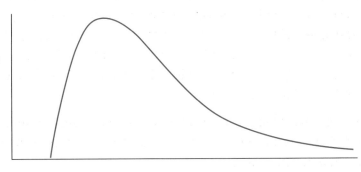

intensity
of radiation
at each
frequency

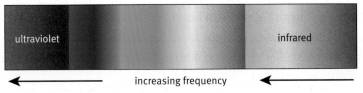

ultraviolet

infrared

← increasing frequency ←

 The graph of the spectrum provides information about intensity as well as showing which frequencies are present.

The Sun's radiation is most intense in the visible part of the spectrum, corresponding to a temperature of about 5500 °C.

The diagram on the right shows the results of comparing the spectra of hotter and cooler stars.

- For a hotter star, the area under the graph is greater; this shows that the luminosity of the star is greater.
- For a hotter star, the **peak frequency** is greater; it produces a greater proportion of radiation of higher frequencies.

These are not special rules for stars; they apply to any hot object.

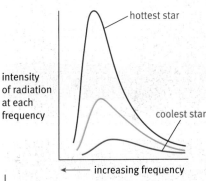

intensity
of radiation
at each
frequency

hottest star

coolest star

← increasing frequency

The spectra of hotter and cooler stars.

Question

1 Stars A and B are the same size, but star A is hotter than star B.
   a Which star has greater luminosity?
   b If you examined the spectra of these stars, which would have the greater peak frequency?
   c Sketch graphs to show how these stars' spectra would differ.

Key words

✓ **electromagnetic spectrum**
✓ **peak frequency**

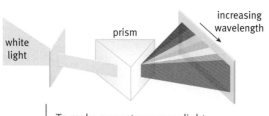

To make a spectrum, pass light through a narrow slit and shine it through a prism.

The mystery of the Sun

It might seem impossible to find out what the Sun is made of. But it turns out you can do this by examining the light it gives out. If sunlight is passed through a prism or diffraction grating, it is split into a spectrum, from red to violet.

In 1802, William Woolaston noticed that the spectrum of sunlight had a strange feature – there were black lines, showing that some wavelengths were missing from the continuous spectrum. These lines are now called Fraunhofer lines, after Joseph von Fraunhofer who made many measurements of their wavelengths.

The colours of the elements

Fraunhofer knew that sodium burns with a yellow flame. When he looked at the spectrum of light from a sodium flame, Fraunhofer saw that it consisted of just a few coloured lines, rather than a continuous band from red to violet.

A spectrum like this is called an **emission spectrum**, because you are looking at the light emitted by a chemical. Today, we know that each element has a different pattern of lines in its emission spectrum, and that this can be used to identify the elements present.

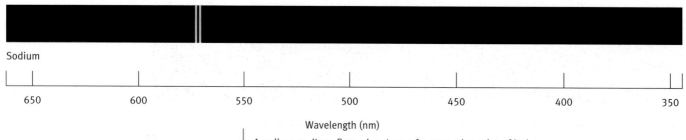

A yellow sodium flame has just a few wavelengths of light.

Making sense of sunlight

The dark lines in the spectrum of sunlight (see opposite) are caused by the absorption of some colours, rather than emission. To understand what is going on, you have to think of the structure of the Sun.

- The surface of the Sun produces white light, with all wavelengths present.
- As this light passes through the Sun's atmosphere, some wavelengths are absorbed by atoms of elements that are present.

As a result, the light that reaches us from the Sun is missing some wavelengths, which correspond to elements in the Sun's atmosphere.

This is an example of an **absorption spectrum**. The wavelengths of the absorption lines reveal which elements have been doing the absorbing. From this, astronomers can identify the elements present in the Sun and in the most distant stars.

Key words

✓ emission spectrum
✓ absorption spectrum

Light from the surface of the Sun must pass through its atmosphere before it reaches us.

Solar atmosphere

photosphere (source of sunlight)

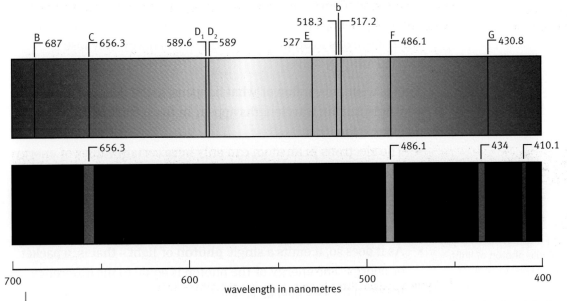

Lines in the emission spectrum of hydrogen (bottom) correspond to some of the dark absorption bands in the Sun's spectrum (top). Other lines in the absorption spectrum are due to other elements in the solar atmosphere, including helium. The numbers are the wavelengths of lines in the hydrogen spectrum in nanometres.

Questions

1 'Light is a messenger from the stars'. Explain how this statement is true.

2 Look at the emission spectrum of hydrogen above. What colours are the main emission lines in this spectrum? Which is the strongest (most intense) line?

A new element

The line spectrum of an element is different from that of every other element – it can be thought of as the 'fingerprint' of the element. In 1868, two scientists used this fact to discover a new element: helium. Norman Lockyer (English) and Jules César Janssen (French) took the opportunity of an eclipse of the Sun to look at the spectrum of light coming from the edge of the Sun. Janssen noticed a line in the spectrum that he had not seen before. He sent his observation to Lockyer, who realised that the line did not correspond to any known element. He guessed that some other element was present in the Sun, and he named it 'helium' after the Greek name for the Sun, 'Helios'. Studies of its spectrum reveal that the Sun is mostly composed of hydrogen and helium, with small amounts of many other elements.

Emitting light

To understand why different elements have different emission spectra, you need to know how atoms emit light. The light is emitted when electrons in atoms lose energy – the energy they lose is carried away by light.

That is a simple version of what happens, and it does not explain why only certain wavelengths appear in the spectrum. Here is a deeper explanation:

- The electrons in an atom can only have certain values of energy. Scientists think of them as occupying points on a 'ladder' of **energy levels**.
- When an electron drops from one energy level to another, it loses energy.
- As it does so, it emits a single **photon** of light – that is, a packet of energy. The energy of the photon is equal to the difference between the two energy levels.

The greater the energy gap, the greater the energy of the photon. High-energy photons correspond to high-frequency, short-wavelength light.

In the simplified diagram of energy levels shown above, you can see that only three photon energies are possible. The most energetic photon comes from an electron that has dropped from the top level to the bottom level.

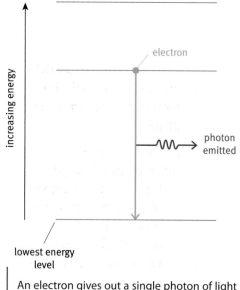

An electron gives out a single photon of light as it drops from one energy level to another.

Absorbing light

The same model can explain absorption spectra. The dark lines come about when electrons absorb energy from white light.

- White light consists of photons with all possible values of energy.
- An electron in a low energy level can only absorb a photon whose energy is just right to lift it up to a higher energy level. When it absorbs such a photon, it jumps to the higher level.
- The white light is now missing photons that have been absorbed because their energies corresponded to the spacings in the ladder of energy levels. The 'missing' photons correspond to the dark lines in an absorption spectrum.

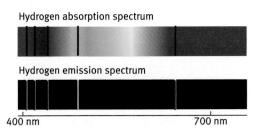

The missing frequencies in the absorption spectrum of hydrogen correspond exactly to the frequencies seen in its emission spectrum.

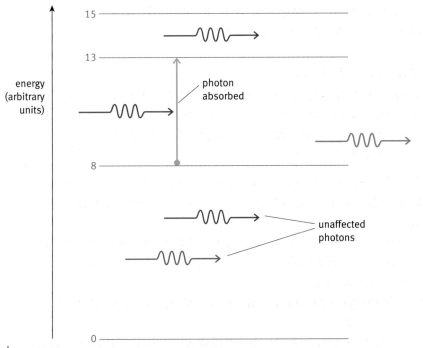

An electron jumps from one energy level to another when the atom absorbs a single photon of light.

The diagram above shows values of energy for each level (in arbitrary units). You can see that:

- Photons of energies 2, 5, 7, 8, 13, and 15 would be absorbed.
- Photons with other energies below 15 would not be absorbed.

A photon with sufficient energy can ionise an atom. An electron gains so much energy that it escapes the attractive force of the nucleus, and leaves the atom. As a result, the atom is left with net positive charge.

Questions

3 Which subatomic particles are associated with the emission and absorption of light?

4 For the energy-level diagram on this page:
 a Explain how photons of energies 2, 5, and 8 would be absorbed.
 b Give other energy values that would also be absorbed.
 c Explain how an atom could emit a photon of energy 13 units.

Key words

- ✔ energy levels
- ✔ photon

The mystery of the Sun

At the end of the 19th century, the source of the Sun's power was a matter of controversy. It had been thought that the Sun was a huge ball of burning coal or some other fuel, or perhaps it was heated by meteorites crashing into its surface. Another idea was that the Sun was shrinking under its own gravity, which would make it heat up.

But geological studies were revealing that the Earth was at least 300 million years old, and that life had evolved over that time. None of the ideas about the Sun's energy source could account for its output of light and heat over such a long time – there was nowhere near enough energy available.

Some influential physicists, including Lord Kelvin, argued that the geologists' and biologists' estimates of the Earth's age were wrong. Their arguments were so forceful that Darwin removed all mention of timescales from his book *On the Origin of Species* (1859).

Mystery solved

The turning point was the discovery of radioactivity and the realisation that radioactive decay could liberate enormous amounts of energy per atom. This led to the idea that nuclear fission and **fusion** reactions could be a source of energy. In 1920, the English astrophysicist Sir Arthur Eddington explained how fusion of hydrogen to produce helium could make the Sun shine for many *thousand* million years.

Nuclear fusion in the Sun

Nuclear fusion occurs when two atomic nuclei get so close together that the **strong nuclear force** makes them react to form a new nucleus. The nuclei's positive charges mean that they repel each other with an electrostatic force. Nuclei can only get close enough to fuse if they approach each other with very high energy. The interior of the Sun is very hot, so the particles are very energetic and nuclear fusion can take place.

Inside the Sun, hydrogen nuclei fuse to make helium. In a sequence of steps, four hydrogen nuclei (protons) make one helium nucleus.

Arthur Eddington, a British astrophysicist, explained how the Sun could be powered by nuclear fusion.

First, two hydrogen nuclei make a deuterium nucleus – the hydrogen isotope that has a proton and a neutron in its nucleus. As the proton decays to a neutron it releases a **positron**, a small positive particle similar to an electron.

$$_1^1H + {}_1^1H \longrightarrow 2\,_1^2H + {}_{+1}^0e^+$$

Notice that the nuclear equation balanced. The sum of the charges is the same both sides, and so is the sum of the atomic numbers. The whole process is shown in the diagram on the right. It can be summarised by the equation:

$$4\,_1^1H \longrightarrow {}_2^4He + 2\,_{+1}^0e^+ + 4\gamma$$

The total mass of the particles after the reaction is less than the total mass beforehand; there is a release of energy. This is described by Einstein's equation:

$$E = mc^2$$

energy released = loss of mass × (speed of light)^2

Overall, fusion of hydrogen to helium involves a mass loss of about 0.7%. Calculations show that hydrogen fusion releases enough energy to provide the Sun's radiation output from the time the Earth formed and for about as long again into the future.

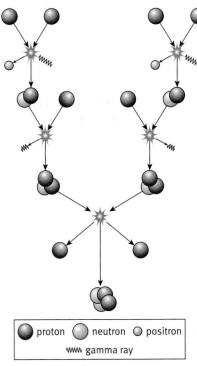

| proton | neutron | positron |

~ gamma ray

Fusion of hydrogen nuclei to make a helium nucleus involves several steps, with two nuclei fusing at each stage. This diagram shows one possible sequence of steps.

Questions

1   The second stage in the sequence shown in the diagram is:

$$_1^1H + {}_1^2H \longrightarrow {}_y^xZ$$

   a   Work out the values of the mass number X and the charge Y of the new nucleus.
   b   Is Z a nucleus of hydrogen or helium?

2   Write a balanced nuclear equation for the final stage shown in the diagram.

3   The Sun contains 1% oxygen nuclei. There are eight protons in an oxygen nucleus. Explain why oxygen nuclei are less likely to fuse together than hydrogen nuclei.

Key words

✔ **fusion**
✔ **strong nuclear force**
✔ **positron**

- ✓ **using a Hertzsprung–Russell diagram to compare stars**
- ✓ **how stars are classified using their temperature and luminosity**
- ✓ **the raw materials for making stars**
- ✓ **the range of conditions found between stars**

Looking at the night sky, you see stars of different brightnesses and colours. By the beginning of the 20th century, astronomers had worked out how to make sense of this:

- Stars might be faint because they were a long way off. Knowing the distance to a faint star, they could work out its luminosity.
- Stars are different colours because they are different temperatures. Red is cool, blue is hot.

A Danish astronomer called Ejnar Hertzsprung set about finding if there was any connection between luminosity and temperature. He gathered together published data and drew up a chart, which he published in 1911.

An American, Henry Russell, came up with the same idea independently. He was unaware of Hertzsprung's chart, which had been published in a technical journal of photography. Today, the chart is known as the Hertzsprung–Russell diagram, or H–R diagram. A modern version is shown below.

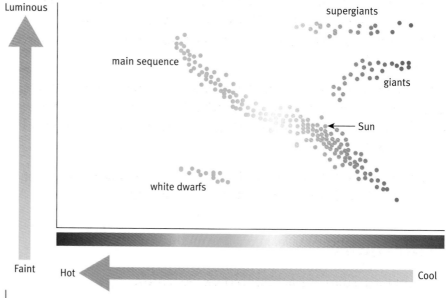

Data for stars – the H–R diagram.

Understanding the H–R diagram

This chart has luminosity on the vertical axis and temperature on the horizontal axis. It is usually drawn as shown, with temperature *decreasing* along the *x*-axis. The Sun is roughly at the middle of the chart.

By 1924, over 200 000 stars had been catalogued. When plotted on the H–R diagram, these stars fell into three groups:

- About 90% of stars (including the Sun) fell along a line running diagonally across the diagram. This is known as the **main sequence**.
- About 10% of stars were **white dwarfs**, small and hot.
- About 1% of stars were **red giants** or **supergiants**, much more luminous than main-sequence stars with the same temperatures.

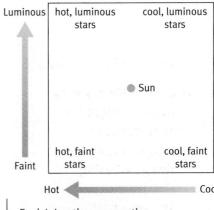

Explaining the axes on the Hertzsprung–Russell diagram.

What the H–R diagram reveals

The stars that appear on the H–R diagram are a representative selection of all stars. A first guess might be that there are simply three different, unrelated types of star. However, astronomers now believe that an individual star changes during its lifetime. They interpret the H–R diagram like this:

- Since most stars are on the main sequence, this suggests that an average star spends most of its lifetime as a main-sequence star.
- A star may spend a small part of its lifetime as a giant, and/or as a white dwarf.

With the exception of supernova explosions, astronomers cannot watch individual stars change through a lifetime because the process is much too slow to see. Instead, they link their observations of star populations to models of how stars work.

In the rest of this section, you will learn how astronomers think stars change, and how this is reflected in the H–R diagram.

Questions

1 List the colours of stars, from coolest to hottest.

2 If the Sun became dimmer and cooler, how would its position on the H–R diagram change?

3 A star in the bottom left-hand corner of the H–R diagram is dim but hot. Why does this suggest that it is small?

Key words

- ✓ **main sequence**
- ✓ **white dwarfs**
- ✓ **red giants**
- ✓ **supergiants**

The Rosette nebula is a hot nebula in the ISM.

What are stars made from?

The 'space' between stars is not empty. It is filled with very low density gas known as the **interstellar medium** (ISM). The gas is mostly hydrogen and helium, with small amounts of other elements. Under certain conditions, some of this gas collects together to make new stars. To tell the story of how stars form, we need to start with the ISM.

The ISM includes clouds of glowing gas such as planetary nebulae, supernova remnants (see page 244) and other hot nebulae. Here the temperature is typically 10 000 °C and there are about 1000 atoms and molecules per cm^3.

Some ISM regions have about a million particles per cm^3 and are sometimes known as 'dense' clouds. But they are not dense by our everyday standards. Normal air has about 10^{20} molecules per cm^3, and even 10^{15} molecules per cm^3 is considered a 'vacuum' on Earth.

'Dense' clouds are very cold, below −200 °C, and are only seen because they block the light from stars or glowing gas behind them.

Between the dense clouds and glowing nebulae is more ISM. Some of this gas is very cold (−100 to −200 °C) with 10−100 particles per cm^3. In other places there is less than one particle per cm^3 and temperatures can reach a million °C.

Questions

4 a Use the information on this page to make a table summarising conditions in various regions of the ISM.

Number of particles per cm^3	Temperature (°C)

b Very roughly, what is the relationship between the numbers in the two columns?

5 Suggest a reason why many hot parts of the ISM glow pink. (Hint: look at page 253.)

Key word
✔ interstellar medium

Great balls of fire

As you have seen, the Sun releases energy by fusing hydrogen nuclei to make helium nuclei. This process takes place in the very hot core of the Sun and most other stars. Between the stars is very thin gas that is mostly very cold. To understand how this gas forms into stars, and how stars get hot enough for nuclear fusion, you need to learn about gases in general.

Describing a gas

Think of a balloon. You blow it up, so it is filled with air. How can you describe the state of that air? What are its properties?

- **Volume** – the amount of space the gas occupies, in m^3.
- **Mass** – the amount of matter, in kg.
- **Pressure** – the force the gas exerts per unit area on the walls of its container, in Pa ($= N/m^2$).
- **Temperature** – how hot the gas is, in °C (or K – see page 264).

These are all measurable quantities that a physicist would use to describe the gas. Understanding how these properties change is important. For example, the engine of a car relies on the pressure of an expanding gas to provide the motive force that makes the car go.

Pressure and volume

Now picture squashing the balloon. You are trying to decrease its volume. Its pressure resists you. It is easier to understand what is happening by pressing on a gas syringe.

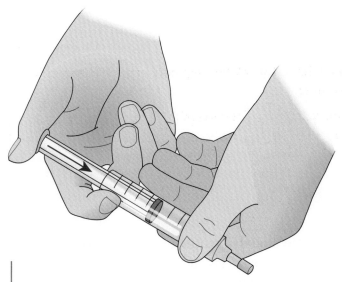

Compressing air in a syringe. This has a closed end so that it contains a fixed mass of air.

Find out about

- ✔ how the pressure, volume, and temperature of a gas are related
- ✔ how ideas about molecules in motion can explain the behaviour of gases
- ✔ the lowest possible temperature
- ✔ how to convert between temperature scales

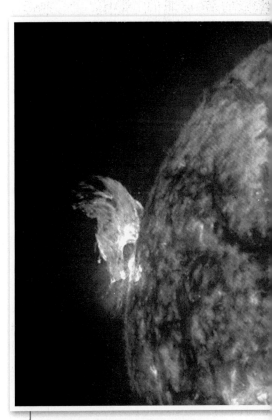

The Sun is a giant ball of hot gas, mostly hydrogen and helium.

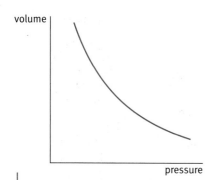

Increasing the pressure on a gas reduces its volume.

- It is easy to push the plunger in a little.
- The more you push on the plunger, the harder it gets to move it further.

This shows that the pressure of the air is increasing as you reduce its volume. If you reduce the force with which you press on the plunger, the air will push back and expand. The graph shows this relationship:

- When the volume of a gas is reduced, its pressure increases.
- The pressure of the gas is inversely proportional to the volume.

For a fixed mass of gas at constant temperature, we can express this relationship as:

$$\text{pressure} \times \text{volume} = \text{constant}$$

Worked example

The value of the constant depends on the amount of gas that you have. For example, suppose you have 100 cm³ of gas in a container at normal atmospheric pressure (1×10^5 Pa). If you squash it with twice atmospheric pressure (2×10^5 Pa), its volume halves to 50 cm³.

Before: pressure × volume = 1×10^5 Pa × 100 cm³ = 100×10^5 (Pa cm³)

After: pressure × volume = 2×10^5 Pa × 50 cm³ = 100×10^5 (Pa cm³)

Explaining pressure

The connection between the pressure and the volume of a gas was worked out before anyone was sure that gases were made of particles.

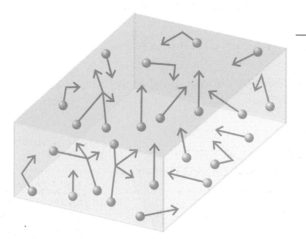

Particles of a gas collide with the walls of its container. This causes pressure.

Yet you can use the **kinetic model** of matter to explain these findings. In this model, a gas consists of particles (atoms or molecules) that move around freely, and most of the volume of the gas is empty space.

- The particles of a gas move around freely. At room temperature, they have speeds around 450 m/s.

Questions

1 Suppose you want to compress 300 cm³ of gas at normal atmospheric pressure into a volume of 100 cm³. How much pressure would you need to apply?

2 An inflated balloon has a volume of 2000 cm³. When the air escapes it has a volume of 4000 cm³ at normal atmospheric pressure. What was the pressure inside the balloon?

- As they move around, they bump into the walls of their container – see the diagram. (They also bump into each other.)
- Each collision with the walls causes a tiny force. Together, billions of collisions produce gas pressure.

Now think about what happens if the same gas is compressed into a smaller volume. The collisions with the walls will be more frequent, and so the pressure will be greater.

As cold as it gets

Now think about what happens when a gas gets cold. Blow up a balloon and put it in the freezer – it starts to shrink. Its pressure and volume have both decreased.

In an experiment to investigate this, it is best not to change one factor (temperature) and then to allow two others (pressure and volume) to both change. So experiments are designed to control one factor while the other is allowed to change. The picture on the right shows how a fixed volume of air (in a rigid flask) can be heated to change its temperature. The gauge shows how the pressure of the gas changes.

Heating up, cooling down

The graphs on the right show the results of experiments like this. Think about the effects of cooling down a fixed mass of gas.

- Fixed volume of gas: as the gas is cooled, its pressure decreases steadily.
- Fixed pressure of gas: as the gas is cooled, its volume decreases steadily.

Both of these graphs show the same pattern: the pressure and volume of the gas decrease as the temperature decreases, and both seem to be heading for a value of zero at a temperature well below 0°C. Whatever gas is used, the graph heads for the same temperature.

The point where a graph like this reaches zero is known as the **absolute zero** of temperature. In practice, all gases condense to form a liquid before they reach this point.

$$\text{absolute zero} = -273°C$$

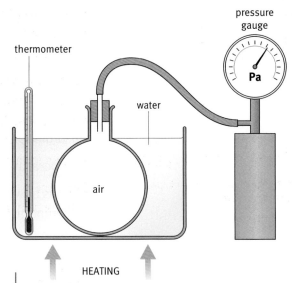

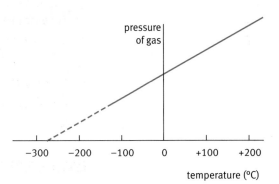

Because the flask is rigid, the volume of the air inside does not change as it is heated or cooled.

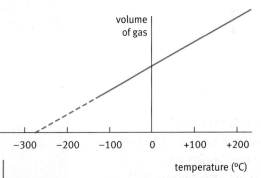

As a gas is cooled, its pressure and volume decrease.

Key words
- ✓ volume
- ✓ mass
- ✓ pressure
- ✓ temperature
- ✓ kinetic model
- ✓ absolute zero

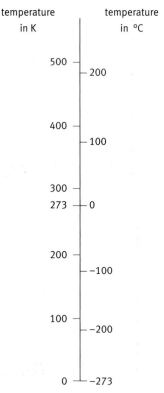

temperature in K | temperature in °C

500
— 200

400
— 100

300
273 — 0

200
— −100

100
— −200

0 — −273

Comparing the Celsius and Kelvin scales of temperature.

Temperature scales

In everyday life, we use the Celsius scale of temperature. This has its zero, 0°C, at the temperature of pure melting ice. You can also define a scale that has its zero at absolute zero, the **Kelvin scale**. The individual divisions on the scale (degrees) are the same as on the Celsius scale, but the starting point is much lower. Because nothing can be colder than absolute zero, there are no negative temperatures on the Kelvin scale.

Temperatures on this scale are given in kelvin (K).

Here is how to convert from one scale to the other:
- temperature in K = temperature in °C + 273
- temperature in °C = temperature in K − 273

So, for example, suppose your body temperature is 37°C. What is this in K?
- Temperature in K = 37°C + 273 = 310 K.

A kinetic explanation

When a gas is cooled down, the particles of the gas lose energy, so they move more slowly.
- If the volume of the gas is fixed, each particle takes longer to reach a wall. So particles strike the walls less frequently. They also strike with less momentum. So the pressure decreases for these two reasons.
- If the pressure is to remain constant, the volume of the gas must decrease to compensate for the fact that the collisions are weaker and less frequent.

Eventually, you can picture the particles of the gas losing all of their kinetic energy, so that they do not collide with the walls at all. There is no pressure. This is absolute zero.

Questions

3 What are the values of the following temperatures on the Kelvin scale?
  a 0°C
  b 100°C
  c −100°C.

4 What are the values of the following temperatures on the Celsius scale?
  a 0 K
  b 8 K
  c 200 K
  d 300 K.

Dark clouds of gas and dust, such as the the Horsehead Nebula in Orion, can have temperatures as low as 8K.

Gas equations

We can redraw the graphs from page 263 using temperature in K. These graphs illustrate two equations that describe how gases behave.

For a fixed mass of gas at constant volume, the pressure is **proportional** to the temperature in kelvin.

$$\frac{\text{pressure}}{\text{temperature (in K)}} = \text{constant}$$

For a fixed mass of gas at constant pressure, the volume is proportional to the temperature in kelvin.

$$\frac{\text{volume}}{\text{temperature (in K)}} = \text{constant}$$

The value of the constant depends on how much gas there is.

For example, if you have a container with 600 cm³ of gas at 300 K, and you heat it to 600 K (keeping it at the same pressure), the volume doubles to 1200 cm³.

$$Before: \frac{\text{volume}}{\text{temperature}} = \frac{600 \text{ cm}^3}{300 \text{ K}} = 2 \text{ cm}^3/\text{K}$$

$$After: \frac{\text{volume}}{\text{temperature}} = \frac{1200 \text{ cm}^3}{600 \text{ K}} = 2 \text{ cm}^3/\text{K}$$

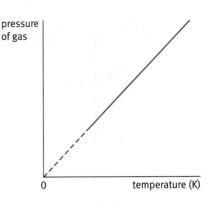

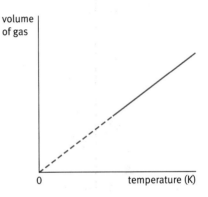

Questions

5 a 1000 cm³ of gas is heated from 300 K to 400 K at constant pressure. What will the new volume be?

   b The gas is now heated from 300 K to 500 K in a container with a fixed volume. At the start its pressure was 1×10^5 Pa. What is its new pressure?

6 Suppose there are two gas clouds in space, and both have the same pressure but cloud A is much hotter than cloud B. What can you say about the density of the two clouds?

7 Imagine that a fixed mass of gas is compressed into half its original volume. The temperature remains constant. Which of the following statements are correct?

   a The pressure of the gas will increase.

   b The average separation of the particles of the gas will decrease.

   c A particle of the gas will strike the walls of the container with greater force.

   d A particle of the gas will strike the walls more frequently.

8 Use the kinetic model of matter to explain why the pressure of a fixed volume of gas increases as the temperature increases.

Key words

✔ **Kelvin scale**

✔ **proportional**

Find out about

- ✔ **how stars and planetary systems form**
- ✔ **the structure of the Sun and other stars**
- ✔ **how energy is transferred within stars**

How does a star form? The raw material of stars is hydrogen and helium, and there has been plenty of that since the early days of the Universe. Here is a simplified version of how astronomers think that stars form:

- A cold cloud of gas and dust starts to contract, pulled together by gravity. It breaks up into several smaller clouds and each continues to contract.
- Within a contracting cloud, each particle attracts every other particle, so that the cloud collapses towards its centre. It forms a rotating, swirling disc.
- As the gas particles are attracted towards the centre, they move faster, which means the gas gets hotter.
- Eventually, the temperature of this material is hot enough for fusion reactions to occur, and a star is born.
- Material further out in the disc clumps together to form planets.

So stars form in **clusters**, and planets form at the same time as the star that they orbit. In these early stages, as the star forms, it is known as a **protostar**. This stage in the Sun's life is thought to have lasted 100 000 to 1 million years.

A protostar at the centre of a new planetary system.

Getting warmer

Here are two ways to think about when a protostar gets hot enough for fusion to start.

- The *gas* idea. The star starts from a cloud of gas. When a gas is compressed, its temperature rises. In this case, the force doing the compressing is gravity.

- The *particle* idea. Every particle in the cloud attracts every other particle. As they 'fall' inwards, they move faster (gravitational potential energy is being converted to kinetic energy). The particles collide with each other, sharing their energy. The fastest particles are at the centre of the cloud (they have fallen furthest), and fast-moving particles mean a high temperature.

Note that these are *not* competing explanations. They are just different ways of describing what is going on.

Seek and find

Astronomers' ideas about star formation are based on what they can observe. For example, they usually observe protostars in clusters close to cold dark clouds (see the picture on page 211). They also use computer models to help test their ideas.

Computer models of star formation can help to explain why the (roughly) spherical material from which a star forms collapses to form a flattened disc. Such models suggest that we will always find that the planets orbiting a star lie in a plane, just as in the Solar System.

Some models of star formation predict that, as a protostar forms, it spins faster and faster. Eventually, it blows out giant jets of hot gas, at right angles to the planetary disc. Large telescopes now have sufficient resolution to allow us to see this going on, as shown in the photo on the right. Planets travelling around distant stars are generally too small to see directly, but the gas jets travel far out into space and can occasionally be spotted.

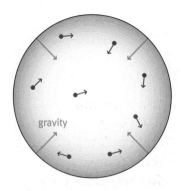

Explaining why a protostar gets hot.

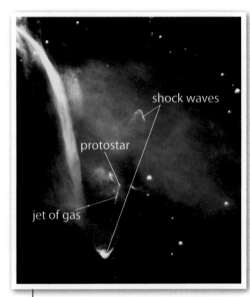

A protostar is forming at the centre of this image. One bright jet of gas can be seen coming downwards from it. Two symmetrical shock waves spread out in opposite directions. (Photo taken by the Very Large Telescope in Chile.)

Questions

1 From what materials does a protostar form?

2 If astronomers see a protostar glowing, does this indicate that nuclear fusion is taking place?

3 Imagine a sky-rocket exploding in the night sky. A small, hot explosion results in material being thrown outwards.
   a Describe the energy changes that are going on.
   b Now imagine the same scene, but in reverse. How is this similar to the formation of a protostar? How does it differ?

4 Suggest reasons why a *cold* cloud of gas is more likely to contract than a hot one.

Key words

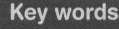

✓ **clusters**
✓ **protostar**

The surface of the Sun, photographed by the *SOHO* satellite, showing the granularity caused by the presence of convective cells.

Main-sequence stars

Spacecraft such as *SOHO* have allowed scientists to look in great detail at the surface of the Sun and to measure the rate at which it is pouring energy out into space. Its colour indicates that the surface temperature of the Sun is about 5800 K – which is far too 'cold' for nuclear fusion reactions to take place.

Modelling the Sun

You cannot tell exactly what is inside the Sun. However, there are some clues that can help physicists to make intelligent guesses:

- Nuclear fusion, the source of the Sun's energy, requires temperatures of millions of degrees.
- Energy leaves the Sun from its surface layer, the **photosphere**, whose temperature is about 5800 K.
- The photosphere has a granular appearance (see the photo), which is continually changing. Something is going on under the surface.
- A star like the Sun can burn steadily for billions of years, so it must radiate energy at the same rate that it generates it from fusion reactions.

Physicists can use these ideas to develop models of the inside of a star. The diagram below shows how they picture the internal structure of the Sun, based on such models.

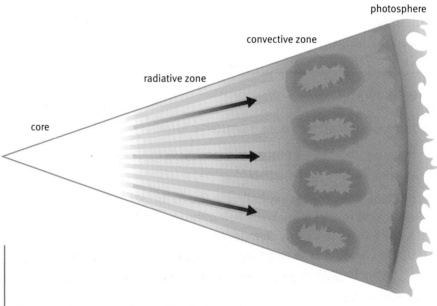

The internal structure of a star like the Sun.

Layer upon layer

- The **core** is the hottest part, with a temperature of the order of 14 million K. This is where nuclear fusion reactions occur. Hydrogen nuclei are fused together to form helium nuclei, releasing energy.

- **Radiation** (photons) travels outwards through the radiative zone.
- Close to the surface, temperatures fall to just 1 million K. Matter can flow quite readily, and **convection** currents are set up, carrying heat energy to the photosphere. This is the convective zone. It is the tops of the convective 'cells' that cause the granular appearance of the Sun's surface.
- Electromagnetic radiation is emitted by the photosphere and radiates outwards through the solar atmosphere.

Other main-sequence stars

Other main-sequence stars can be modelled in the same way as the Sun. All main-sequence stars are fusing hydrogen in their cores to make helium. A main-sequence star has a steady luminosity and temperature for all the time that it is fusing hydrogen in its core – millions, or even billions, of years.

Differences between main-sequence stars are due to their different masses. The more massive the star, the hotter its core and the more rapidly it turns hydrogen into helium. The most massive main-sequence stars are also the hottest and the most luminous. The table below lists some typical values.

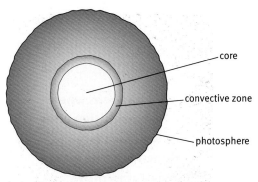

Cross-section of a star five times the mass of the Sun.

Mass of star	Luminosity	Surface temperature (K)
0.5 × Sun	0.03 × Sun	3800
1 × Sun	1 × Sun	5800
3 × Sun	60 × Sun	11 000
15 × Sun	17 000 × Sun	28 000

The lifetime of a star on the main sequence depends on its mass and the rate at which it turns hydrogen into helium.

Questions

5 Why does hydrogen fusion take place only in the *core* of a main-sequence star?

6 On a sketch copy of an H–R diagram (see page 258), label the ends of the main sequence to show the most massive and the least massive stars.

7 The helium made by a main-sequence star stays in its core. Look at the cross-section diagrams on these pages and put forward a reason for this.

8 The greater the mass of a star, the *shorter* the time it spends on the main sequence. Suggest an explanation for this.

Key words

- ✔ **photosphere**
- ✔ **core**
- ✔ **radiation**
- ✔ **convection**

Many generations into the future, people can expect the Sun to keep releasing energy at a steady rate. Fusion reactions will continue in its core, as hydrogen is converted to helium.

But this cannot go on for ever, because eventually all of the hydrogen in the Sun's core will be used up. What happens then?

As fusion slows down in the core of any star, its core cools down and there is less pressure, so the core collapses. The star's outer layers, which contain hydrogen, fall inwards, becoming hot. This causes new fusion reactions, making the outer shell expand. At the same time, the surface temperature falls, so that the colour changes from yellow to red. This produces a red giant.

In the case of the Sun, calculations suggest that it may expand sufficiently to engulf the three nearest planets – Mercury, Venus, and Earth.

An artist's impression of the view from a planet when its star has become a red giant. A moon is also shown, for comparison.

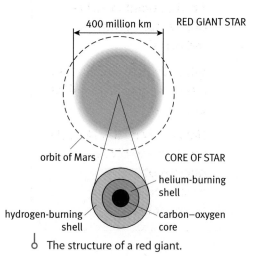

400 million km — RED GIANT STAR

orbit of Mars — CORE OF STAR

helium-burning shell

hydrogen-burning shell — carbon–oxygen core

The structure of a red giant.

Inside a red giant

While the outer layers of a red giant star are expanding, its core is contracting and heating up to 100 million K. This is hot enough for new fusion reactions to start. Helium nuclei have a bigger positive charge than hydrogen nuclei, so there is greater electrical repulsion between them. If they are to fuse, they need greater energy to overcome this repulsion. When helium nuclei do fuse, they form heavier elements such as carbon, nitrogen, and oxygen, releasing energy.

After a relatively short period (a few million years), the outer layers cool and drift off into space. The collapsed inner core remains as a white dwarf. No fusion occurs in a white dwarf so it gradually cools and fades.

Life of the Sun

Picture the life of a star like the Sun on the Hertzsprung–Russell diagram (page 258).

- Protostars are to the right of the main sequence. As it heats up, a protostar moves to a point on the main sequence, where it stays for billions of years.
- When it becomes a red giant, it moves above the main sequence.
- Finally, as a white dwarf, it appears below the main sequence.

More massive stars

The Sun is a relatively small star. Its core won't get hot enough to fuse elements beyond carbon. Bigger stars, greater than about 8 solar masses, also expand, to become supergiants. In these, core temperatures may exceed 3 billion degrees and more complex fusion reactions can occur, forming even heavier elements and releasing yet more energy. But this cannot go on forever; even massive stars do not make elements heavier than iron.

Beyond iron

When a star gets as far as making iron in its core, events take a dramatic turn.

So far, fusion of nuclei to make heavier ones has involved a release of energy. The star's core is heated, raising the pressure and stopping the star collapsing under its own gravity.

But when nuclei heavier than iron are made by fusing lighter nuclei, there is an overall increase in mass. This means that some input of energy is needed (remember: $E = mc^2$), rather than energy being released. Once iron is made in a star's core, there is no further release of energy.

Supernova explosion

What happens after the supergiant phase of a massive star is one of the most dramatic events in nature. A star of about 8 solar masses or more can get as far as making iron in its core. Iron nuclei absorb energy when they fuse, and there is no source of heating to keep up the pressure in the core. Now the drama starts.

The outer layers of the star are no longer held up by the pressure of the core, and they collapse inwards. The core has become very dense, and the outer material collides with the core and bounces off, flying outwards. The result is a huge explosion called a supernova.

> ## Questions
>
> 1 At what point in its life does a star become a red giant?
>
> 2 What determines whether a star becomes a red giant or a supergiant?
>
> 3 How many helium-4 nuclei must fuse to give a nucleus of:
>   a carbon-12?
>   b oxygen-16?

In the course of the explosion, temperatures rise to 10 billion K, enough to cause the fusion of medium-weight elements and thus form the heaviest elements of all – up to uranium in the periodic table. For a few days, a supernova can outshine a whole galaxy.

The remnants of a supernova in the constellation of Cassiopeia. The cloud is about 5 parsecs across. This is a composite image, made using three telescopes to capture infrared, visible, and X-ray data.

Questions

4 On a sketch copy of an H–R diagram (page 258), draw and label a line tracing out the life of a Sun-like star from protostar to white dwarf.

5 Put these objects in order, from least dense to most dense:
neutron star, protostar, supergiant, black hole, main-sequence star

The next generation

The photograph above shows the remnants of a supernova that happened in about 1660. You can see the expanding sphere of dust and gas, formed from the star's outer layers. This material contains all of the elements of the periodic table. As it becomes distributed through space, it may become part of another contracting cloud of dust and gas. A protostar may form with new planets orbiting it, and the cycle starts over again.

Dense and denser

The core of an exploding supernova remains. If its mass is less than about 2.5 solar masses, this central remnant becomes a **neutron star**. This is made almost entirely of neutrons, compressed together like a giant atomic nucleus, perhaps 30 km across.

A more massive remnant collapses even further under the pull of its own gravity, to become a **black hole**. Within a black hole, the pull of gravity is so strong that not even light can escape from it.

Neutron stars are thought to explain the pulsars, discovered by Jocelyn Bell and Anthony Hewish (page 210). As the core of a star collapses to form a neutron star, it spins faster and faster. Its magnetic field becomes concentrated, and this results in a beam of radio waves coming out of its magnetic poles. As the neutron star spins round, this beam sweeps across space and might be detected as a regular series of pulses at an observatory on some small, distant planet.

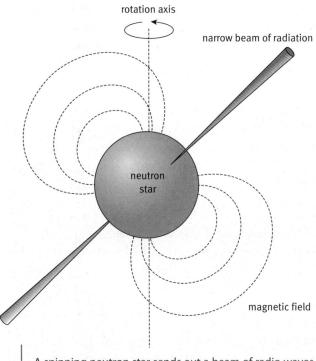

A spinning neutron star sends out a beam of radio waves – the origin of a pulsar.

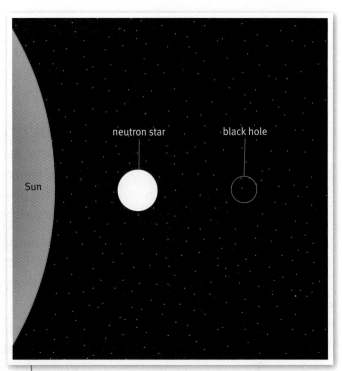

The neutron star and the black hole shown here have the same mass as the Sun.

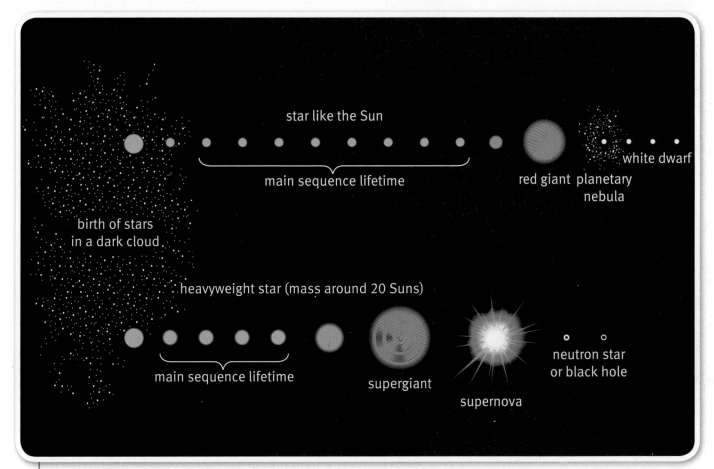

The life and death of a star mainly depends on its mass. The most massive stars have the shortest lives and most spectacular deaths.

This module started with people making simple naked-eye observations of the sky, as they have done for thousands of years. It ends with a simple question about our place in the Universe – are we alone?

Life on other planets?

The story of star formation (pages 266–269) also explains how planets might be formed. Some of the gas and dust in a collapsing cloud orbits around the protostar rather than falling into it. Over time, this gas and dust gradually clumps together to make planets.

If the gas and dust includes some remnants of supernova explosions, then it will contain traces of elements from the whole periodic table, including those (such as carbon and oxygen) needed for life as we know it on Earth.

So there is an intriguing argument that goes like this:

- There are about 100 thousand million stars in the Milky Way, so it seems likely that at least some of them have planets.
- If there are planets round other stars, then maybe life has evolved on some of them. It might be possible to detect some living organisms by the effect they have on a planet.
- If life is present, some organisms might be intelligent. It might be possible to communicate with intelligent organisms on other planets. They might be sending out signals. Or they might have detected our presence on Earth.

Exoplanets

Towards the end of the 20th century, astronomers started to look for planets orbiting other stars, known as **exoplanets**. This search needs very sensitive telescopes with good resolving power. Between 1990 and 2010 astronomers found over 460 exoplanets and they are still finding more.

Astronomers detect exoplanets by clever techniques such as small dips in the brightness of a star as its planet passes in front of it, or by the wobbling motion of the star caused by the gravity of a planet. The image opposite shows a planet orbiting a star that is surrounded by a disc of dust. The star is about 25 light-years away.

So far, all exoplanets detected have been much more massive than Earth.

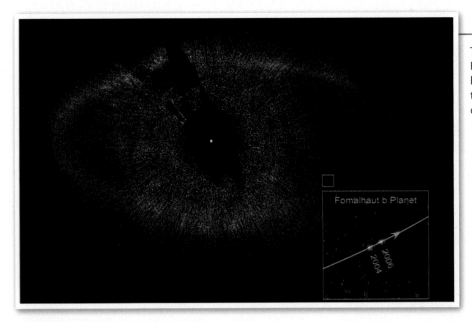

This image shows the exoplanet known as Fomalhaut b in 2004 and 2006. Astronomers have calculated that it takes 872 years to orbit the star, and its orbit is about 115 times the size of Earth's orbit.

SETI

As long ago as the 1950s, when the first satellites were launched into Earth orbit, scientists had the idea that it might be possible to detect radio signals sent by intelligent life elsewhere in the Universe. This marked the start of the Search for Extra-Terrestrial Intelligence (**SETI**). Since then, several SETI projects have taken place. One of the biggest is the SETI @home project, where some 50 000 people around the world use their home computers to process data from radio telescopes to see whether it contains any sign of 'intelligent' signals.

So far, there is no evidence of life elsewhere.

This vast radio telescope in Arecibo, Puerto Rico, has been used to scan the sky for evidence of 'intelligent' radio signals.

Questions

1 There may be intelligent life forms on exoplanets. What risks and benefits could there be in communicating with them?

2 The closest stars to our Sun are several light-years away. What problems might this cause if we wanted to communicate with life forms on planets orbiting other stars?

3 Suppose you are an alien on a planet 10 light-years from Earth. Describe any possible evidence you could have to suggest that the Earth exists and there is life on it.

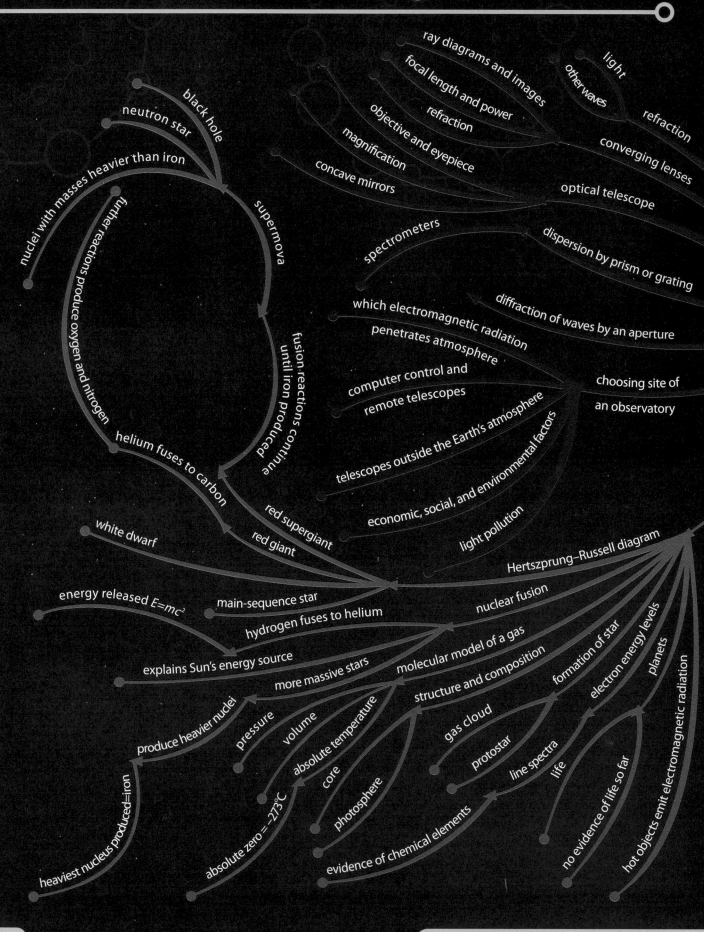

ray diagrams and images

focal length and power

light

other waves

refraction

refraction

objective and eyepiece

converging lenses

magnification

concave mirrors

optical telescope

black hole

neutron star

supernova

spectrometers

dispersion by prism or grating

nuclei with masses heavier than iron

further reactions produce oxygen and nitrogen

which electromagnetic radiation penetrates atmosphere

diffraction of waves by an aperture

computer control and remote telescopes

choosing site of an observatory

fusion reactions continue until iron produced

telescopes outside the Earth's atmosphere

economic, social, and environmental factors

helium fuses to carbon

red supergiant

light pollution

white dwarf

red giant

Hertszprung–Russell diagram

energy released $E=mc^2$

main-sequence star

nuclear fusion

hydrogen fuses to helium

molecular model of a gas

explains Sun's energy source

more massive stars

structure and composition

formation of star

electron energy levels

planets

produce heavier nuclei

pressure

volume

absolute temperature

gas cloud

protostar

line spectra

life

no evidence of life so far

hot objects emit electromagnetic radiation

heaviest nucleus produced=iron

absolute zero = −273°C

core

photosphere

evidence of chemical elements

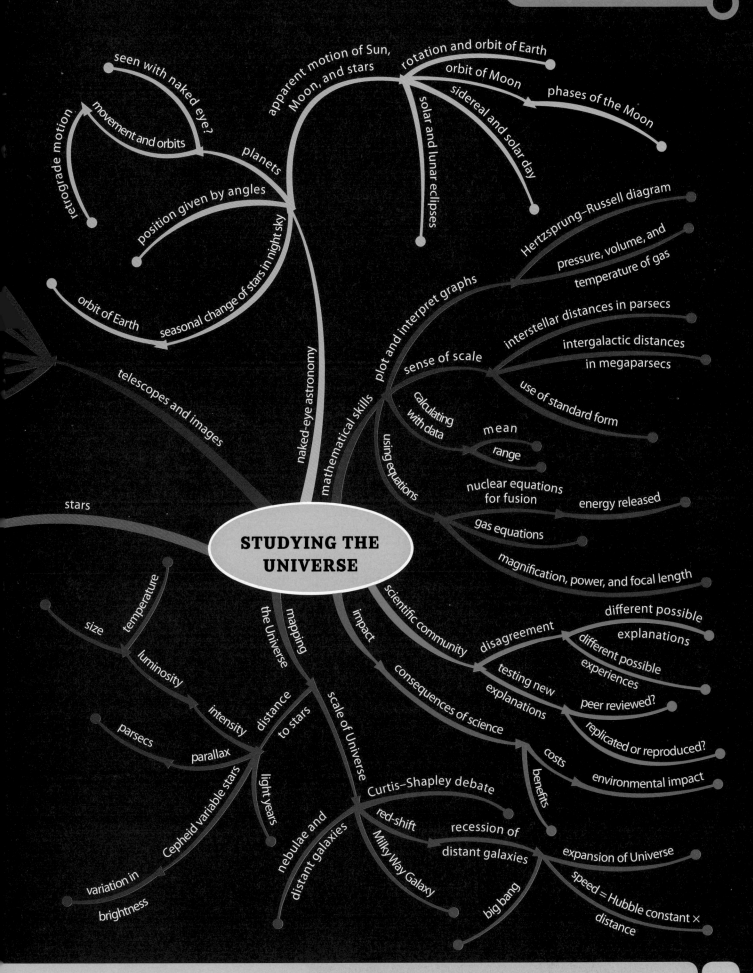

Science
Explanations

Astronomy is a very ancient science. People have been observing and tracking the stars and planets for many hundreds of years to learn more about our Universe.

You should know:
- how the Sun, Moon, and stars appear to travel across the night sky
- how and why a sidereal day is different to a solar day
- about the way the Sun, Moon, planets, and stars appear to move because of their orbits and the rotation of the Earth
- how to explain, using the position of the Sun, Moon, and Earth:
  - the phases of the Moon
  - solar and lunar eclipses
- that the difference in tilt of the Earth's orbit around the Sun and the Moon's orbit around the Earth explains why eclipses do not occur more frequently
- that the position of astronomical objects seen from Earth is measured by angles
- that the different stars in the night sky at different times of year is explained by the movement of the Earth around the Sun
- which planets can be seen with the naked eye and that all the planets move against the background of stars
- that planets sometimes appear to move with retrograde motion (in the opposite direction to the stars) and how to explain why this happens.

Telescopes and images

A telescope makes distant objects appear larger and closer, so that more detail can be seen. It also allows fainter objects to be studied because a telescope has a larger aperture than the human eye and collects more light. Modern telescopes can produce lasting images using electronic detectors or photographic film.

You should know:
- about the refraction of light and other waves
- how converging (convex) lenses use refraction to form an image
- how to draw and interpret ray diagrams that show how converging lenses form images
- about the focal length and power of a lens and how to calculate them
- that an optical telescope has an objective lens, or mirror, and an eyepiece

- about optical telescopes, including how to calculate the magnification
- about dispersion of light by a prism, a grating, and spectrometers
- why most astronomical telescopes use mirrors
- how concave mirrors bring light to a focus
- about the diffraction of waves by an aperture, and how this affects telescopes
- which electromagnetic radiation can get through the atmosphere
- the effect of light pollution on astronomy
- the locations of major optical and infrared observatories on Earth
- the factors that influence the choice of site for major observatories
- how astronomers work with local and remote telescopes
- the advantages of computer-controlled telescopes
- the advantages and disadvantages of telescopes outside the Earth's atmosphere
- the advantages to the astronomy community of working together
- the economic, environmental, and social factors involved in planning, building, operating, and closing down an observatory.

Mapping the Universe

In order to understand and explain the Universe and the objects we see in it, we need to know how far away they are. Once an object's distance is known, then astronomers can work out other things, such as how big it is and how much energy it is giving out.

You should know:

- about the parallax effect and how it can be used to measure the distance to stars
- that the parsec and light-year are units of distance, and how they are defined
- that the luminosity of a star depends on its temperature and size
- why the intensity of light from a star depends on its luminosity and distance from Earth
- about Cepheid variable stars, how they vary in brightness, and how they can be used to measure distance from Earth
- about the role of Cepheid variable stars in showing the scale of the Universe and that most spiral nebulae are distant galaxies
- about the Curtis–Shapley debate
- that the Sun is a star in the Milky Way galaxy
- about Hubble's discovery of distant galaxies and about the red-shift in the electromagnetic radiation received from distant galaxies
- that the motion of galaxies suggests the Universe is expanding
- about the scale of the Universe, including that interstellar distances are measured in parsecs, and intergalactic distances in megaparsecs
- how to use the equation speed of recession = Hubble constant × distance
- that scientists believe the Universe began with a 'big bang' about 14 thousand million years ago.

Stars

Scientists study the star at the centre of the Solar System, the Sun, to understand more about other stars and about the Universe. The radiation produced by stars reveals their structure and composition and allows us to find out how they are formed and what will happen to them at the end of their life.

You should know:

- that hot objects emit a continuous range of electromagnetic radiation with luminosity and peak frequency that depends on temperature
- about electron energy levels in atoms and how these give rise to line spectra
- that specific spectral lines in the spectrum of a star provide evidence of the chemical elements present in it and how to use data to identify elements
- how, in the early 20th century, nuclear fusion explained the Sun's energy source, which, until then, had been a mystery
- that the nuclear fusion of hydrogen to form helium releases energy
- how to complete nuclear equations relating to fusion in stars (including positron emission) and be able to calculate the energy released using $E = mc^2$
- about the structure and composition of stars
- about the different stars found in different regions of the Hertzsprung–Russell diagram, which is a plot of star temperature against luminosity
- that more massive stars have hotter cores and create heavier nuclei, up to the mass of the iron nucleus
- how to explain, using a molecular model, that the volume of a gas is inversely proportional to its pressure at constant temperature
- that the absolute zero of temperature is –273ºC, which is 0 K, and how to convert between temperatures in ºC and temperatures in kelvin
- that gas pressure and volume are proportional to the absolute temperature
- how to calculate pressure, volume, and temperature using the gas equations
- how to explain the formation of a protostar, and a star, from a cloud of gas
- how to explain what happens to a star when its hydrogen runs out, including the formation of red giants, supergiants, white dwarfs, and supernovae
- that in red giants helium fuses to form carbon, followed by further reactions that produce heavier nuclei such as nitrogen and oxygen
- how to explain that supernovae form nuclei with masses heavier than iron and leave a neutron star or black hole
- that astronomers have evidence of hundreds of planets around nearby stars
- why many scientists think that life exists elsewhere in the Universe
- that no evidence of extraterrestrial life in the present or past has been detected.

Ideas about Science

In addition to developing an understanding of the composition of the Universe, it is important to understand how the scientific community works together and how decisions are made about science and technology.

Scientists use their knowledge and evidence to suggest explanations for their observations and discoveries. They share their results with the scientific community to see if other scientists agree with the explanations. You should be able to:

* describe the peer-review process in which new scientific claims are evaluated by other scientists; the Curtis–Shapley debate was an example of this; both Heber D. Curtis and Harlow Shapley presented their evidence to other astronomers
* recognise that new scientific claims that have not yet been evaluated by the scientific community are less reliable than well-established ones.
* identify the fact that a finding has not been reproduced by another scientist as a reason for questioning a scientific claim
* explain that scientists think it is important for results to be replicated, because if the same results are obtained by other scientists who are completely independent this rules out lots of reasons for doubting the results, for example, errors in technique, poor equipment, or a bias on the part of those doing the experiment
* show that you are aware that the same data may be interpreted in more than one way; for example, the data that showed galaxies were moving apart was interpreted by some scientists as showing that the Universe started from a small point – the big bang theory – but other scientists thought the data supported the steady-state theory in which galaxies moved apart and new ones appeared in the gaps
* suggest reasons why scientists may disagree, for example, their personal background, experience, or interests may influence their judgement
* discuss what may happen when new data disagrees with predictions using an accepted explanation; for example, at the end of the 19th century geological data showed the Earth to be at least 300 million years old – however, none of the possible accepted explanations for the Sun's energy source predicted that the Sun could last this long, so the age for the Earth was not accepted until a new theory for the energy source of the Sun – nuclear fusion – was suggested
* suggest reasons why scientists should not give up an accepted explanation immediately if new data appears to conflict with it; there may be another explanation. When the planet Uranus was observed to be in a different position from that predicted by its orbit around the Sun, instead of saying that Newton's laws of motion did not apply to Uranus, it was suggested that this might be due to the presence of another planet. This was later confirmed when Neptune was discovered.

In making decisions about science and technology, you should be able to:
* identify the groups affected and the main benefits and costs of a course of action for each group; for example, when an observatory is built in a remote location the main groups affected will be local people, the astronomers who may visit, and other people who may move to the area for jobs – you should be able to suggest benefits and costs for these groups
* suggest reasons why different decisions about choosing the site for an observatory might be appropriate in view of differences in social and economic context.

Review Questions

1 A star chart shows what can be seen in the night sky for a particular time and date and observation position on the Earth.

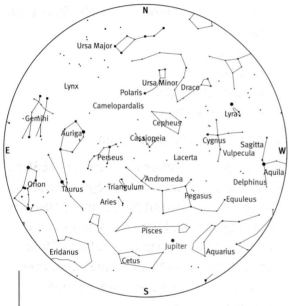

Star chart for 18th January 2011 at 18:00, observed from York, UK.

Describe and explain why the observations would look different from the chart shown for:

a observations of the stars made from York two hours later

b observations of Jupiter made from York two hours later

c observations of the stars made from York six months later

d observations of Jupiter one year later

e observations of the stars made from Australia on the same date.

2 Anna wants to make a telescope. The table shows some properties of the lenses that are available.

Lens	Diameter in mm	Focal length in mm
A	16.5	30
B	16.5	49
C	16.5	65
D	34.5	65
E	34.5	225

a Calculate the power in dioptres of lens **A**.

b Which lens would be the best lens for the eyepiece of a telescope? Explain your answer.

c Which lens would be the best lens for the objective lens? Explain your answer.

d Most astronomical telescopes have a concave mirror as the objective. Explain why a mirror is used instead of a lens.

3 Astronomers use various methods of measuring distances to stars.

a Explain why the parallax method is less precise for more distant stars.

b Explain why there is a problem with using the observed brightness of a star to estimate its distance.

c Explain the property of Cepheid variable stars that is used to measure their distance.

4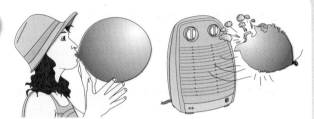

A balloon is inflated and a knot tied in the end to seal it. The balloon is placed near a heater and soon bursts.

Use the molecular model of a gas to explain why the balloon bursts.

5 A car tyre is inflated to a pressure of 200 kPa on a cold morning when the temperature is 7°C. Later in the day after a long car journey the temperature of the tyre rises to 27°C.

a Convert the two temperatures to values in K.

b Calculate the pressure of the air in the tyre when the temperature has risen to 27 °C.

c What assumption did you have to make to carry out your calculation?

6 The Hertzsprung–Russell diagram shows patterns in the properties of stars.

temperature

a Copy the axes and add labels to the axes:
  - faint
  - hot
  - luminous
  - cool

b Show the regions on the graph where each star appears:
  - supergiant
  - giant
  - main sequence
  - white dwarf

c Add the Sun to the diagram.

7 The European Southern Observatory (ESO) is a joint project between 14 European countries. The project carries out astronomical research using large optical telescopes placed in the Atacama desert high in the Andes in Chile. The headquarters of the organisation, where many of the scientists are based, are in Munich, Germany.

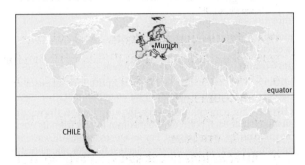

a Suggest some of the advantages of European countries carrying out a joint research project.

b Suggest and explain the advantages of having the headquarters of the organisation in Germany, but the telescopes in the mountains of Chile.

Glossary

absolute zero Extrapolating from the behaviour of gases at different temperatures, the theoretically lowest possible temperature, –273°C. In practice, the lowest temperature achievable is about a degree above this.

absorb (radiation) The radiation that hits an object and is not reflected, or transmitted through it, is absorbed (for example, black paper absorbs light). Its energy makes the object get a little hotter.

absorption spectrum (of a star) Consists of dark lines superimposed on a continuous spectrum. It is created when light from the star passes through a cooler gas that absorbs photons of particular energies.

acceleration The rate of change of an object's velocity, that is, its change of velocity per second. In situations where the direction of motion is not important, the change of speed per second tells you the acceleration.

action at a distance An interaction between two objects that are not in contact, where each exerts a force on the other. Examples include two magnets, two electric charges, or two masses (for example, the Earth and the Moon).

activity The rate at which nuclei in a sample of radioactive material decay and give out alpha, beta, or gamma radiation.

actual risk Risk calculated from reliable data.

aerial A wire, or arrangement of wires, that emits radio waves when there is an alternating current in it, and in which an alternating current is induced by passing radio waves. So it acts as a source or a receiver of radio waves.

air resistance The force exerted on an object by the air, when it moves through it. Its direction is opposite to the direction in which the object is moving. It depends on the speed of the moving object (it increases as the object gets faster) and on its size and shape (as it is caused by the object having to move the air in front of it aside as it goes).

alpha radiation The least penetrating type of ionizing radiation, produced by the nucleus of an atom in radioactive decay. A high-speed helium nucleus.

alternating current (a.c.) An electric current that reverses direction many times a second.

ammeter A meter that measures the size of an electric current in a circuit.

ampere (amp) The unit of electric current.

amplifier A device for increasing the amplitude of an electrical signal. Used in radios and other audio equipment.

amplitude For a mechanical wave, the amplitude is the maximum distance that each point on the medium moves from its normal position as the wave passes. For an electromagnetic wave, it is the maximum value of the varying electric field (or magnetic field).

analogue signal Signal used in communications in which the amplitude can vary continuously.

angular magnification (of a refracting telescope) The ratio of the angle subtended by an object when seen through the telescope to the angle subtended by the same object when seen with the naked eye. It can be calculated as focal length of objective lens / focal length of eyepiece lens.

aperture (of a telescope) The light-gathering area of the objective lens or mirror.

asteroid A dwarf rocky planet, generally orbiting the Sun between the orbits of Mars and Jupiter.

astrolabe An instrument used for locating and predicting the positions of the Sun, Moon, planets, and stars, as well as navigating and telling the time.

atmosphere The Earth's atmosphere is the layer of gases that surrounds the planet. It contains roughly 78% nitrogen and 21% oxygen, with trace amounts of other gases. The atmosphere protects life on Earth by absorbing ultraviolet solar radiation and reducing temperature extremes between day and night.

attract Pull towards.

average speed The distance moved by an object divided by the time taken for this to happen.

background radiation The low-level radiation, mostly from natural sources, that everyone is exposed to all the time, everywhere.

beta radiation One of several types of ionising radiation, produced by the nucleus of an atom in radioactive decay. More penetrating than alpha radiation but less penetrating than gamma radiation. A high-speed electron.

big bang An explosion of a single mass of material. This is currently the accepted scientific explanation for the start of the Universe.

black hole A mass so great that its gravity prevents anything escaping from it, including light. Some black holes are the collapsed remnants of massive stars.

carbon cycle The human and natural processes that move carbon and carbon compounds continuously between the Earth, its oceans and atmosphere, and living things.

carrier A steady stream of radio waves produced by an RF oscillator in a radio to carry information.

Cepheid variable A star whose brightness varies regularly, over a period of days.

chain reaction A process in which the products of one nuclear reaction cause further nuclear reactions to happen, so that more and more reactions occur and more and more product is formed. Depending on how this process is controlled, it can be used in nuclear weapons or the nuclear reactors in power stations.

charged Carrying an electric charge. Some objects (such as electrons and protons) are permanently charged. A plastic object can be charged by rubbing it. This transfers electrons to or from it.

climate Average weather in a region over many years.

coding (in communications) Converting information from one form to another, for example, changing an analogue signal into a digital one.

comet A rocky lump, held together by frozen gases and water, that orbits the Sun.

commutator An arrangement for changing the direction of the electric current through the coil of a motor every half turn. It consists of a ring divided into two halves (a split ring) with two contacts (called brushes) touching the two halves. It also allows the coil to turn continuously without the connecting wires getting tangled up.

compression A material is in compression when forces are trying to push it together and make it smaller.

conservation of energy The principle that the total amount of energy at the end of any process is always equal to the total amount of energy at the beginning – though it may now be stored in different ways and in different places.

constellation A group of stars that form a pattern in the night sky. Patterns recognised are cultural and historical, and are not based on the actual positions of the stars in space.

contamination (radioactive) Having a radioactive material inside the body, or having it on the skin or clothes.

control rod In a nuclear reactor, rods made of a special material that absorbs neutrons are raised and lowered to control the rate of fission reactions.

convective zone (of a star) The layer of a star above its radiative zone, where energy is transferred by convective currents in the plasma.

temperature remains the same. Ohm's law does not apply to all conductors.

optical fibre Thin glass fibre down which a light beam can travel. The beam is reflected at the sides by total internal reflection, so very little escapes. Used in modern communications, for example, to link computers in a building into a network.

ozone layer A thin layer in the atmosphere, about 30 km up, where oxygen is in the form of ozone molecules. The ozone layer absorbs ultraviolet radiation from sunlight.

parallax The apparent shift of an object against a more distant background, as the position of the observer changes. The further away an object is, the less it appears to shift. This can be used to measure how far away an object is, for example, to measure the distance to stars.

parallax angle When observed at an interval of six months, a star will appear to move against the background of much more distant stars. Half of its apparent angular motion is called its parallax angle.

parsec (pc) A unit of astronomical distance, defined as the distance of a star that has a parallax angle of one arcsecond. Equivalent to 3.1 3 1012 km.

peak frequency The frequency with the greatest intensity.

peer review The process whereby scientists who are experts in their field critically evaluate a scientific paper or idea before and after publication.

penumbra An area of partial darkness in a shadow, for example, places in the Moon's path where the Earth only partially blocks off sunlight. Some sunlight still reaches these places because the Sun has such a large diameter.

phases (of the Moon) Changing appearance, due to the relative positions of the Earth, Sun, and Moon.

photon Tiny 'packet' of electromagnetic radiation. All electromagnetic waves are emitted and absorbed as photons. The energy of a photon is proportional to the frequency of the radiation.

photosphere The visible surface of a star, which emits electromagnetic radiation.

planet A very large, spherical object that orbits the Sun, or other star.

positive A label used to name one type of charge, or one terminal of a battery. It is the opposite of negative.

potential difference (p.d.) The difference in potential energy (for each unit of charge flowing) between any two points in an electric circuit. Also called voltage.

power In an electric circuit, the rate at which work is done by the battery or power supply on the components in a circuit. Power is equal to current × voltage.

pressure (of a gas) The force a gas exerts per unit area on the walls of its container.

primary energy source A source of energy not derived from any other energy source, for example, fossil fuels or uranium.

principal axis An imaginary line perpendicular to the centre of a lens or mirror surface.

proportional Two variables are proportional if there is a constant ratio between them.

proton A positively charged particle found in the nucleus of atoms. The relative mass of a proton is 1 and it has one unit of charge.

protostar The early stages in the formation of a new star, before the onset of nuclear fusion in the core.

radiation A flow of information and energy from a source. Light and infrared are examples. Radiation spreads out from its source, and may be absorbed or reflected by objects in its path. It may also go (be transmitted) through them.

radiation dose A measure, in millisieverts, of the possible harm done to your body, which takes into account both the amount and type of radiation you are exposed to.

radiative zone (of a star) The layer of a star surrounding its core, where energy is transferred by photons to the convective zone.

radio waves Electromagnetic waves of a much lower frequency than visible light. They can be made to carry signals and are widely used for communications.

radioactive Used to describe a material, atom, or element that produces alpha, beta, or gamma radiation.

radioactive dating Estimating the age of an object such as a rock by measuring its radioactivity. Activity falls with time, in a way that is well understood.

radioactive decay The spontaneous change in an unstable element, giving out alpha, beta, or gamma radiation. Alpha and beta emission result in a new element.

radiotherapy Using radiation to treat a patient.

ray diagram A way of representing how a lens or telescope affects the light that it gathers, by drawing the rays (which can be thought of as very narrow beams of light) as straight lines.

reaction (of a surface) The force exerted by a hard surface on an object that presses on it.

reflection What happens when a wave hits a barrier and bounces back off it. If you draw a line at right angles to the barrier, the reflected wave has the same angle to this line as the incoming wave. For example, light is reflected by a mirror.

reflector A telescope that has a mirror as its objective. Also called a reflecting telescope.

refraction Waves change their wavelength if they travel from one medium to another in which their speed is different. For example, when travelling into shallower water, waves have a smaller wavelength as they slow down.

refractor A telescope that has a lens as its objective, rather than a mirror.

renewable energy source Resources that can be used to generate electricity without being used up, such as the wind, tides, and sunlight.

repel Push apart.

resistance The resistance of a component in an electric circuit indicates how easy or difficult it is to move charges through it.

resolving power The ability of a telescope to measure the angular separation of different points in the object that is being viewed. Resolving power is limited by diffraction of the electromagnetic waves being collected.

resultant force The sum, taking their directions into account, of all the forces acting on an object.

retrograde motion An apparent reversal in a planet's usual direction of motion, as seen from the Earth against the background of fixed stars. This happens periodically with all planets beyond the Earth's orbit.

risk The probability of an outcome that is seen as undesirable, associated with some behaviour or process.

rock cycle Continuing changes in rock material, caused by processes such as erosion, sedimentation, compression, and heating.

sampling In the context of physics, measuring the amplitude of an analogue signal many times a second in order to convert it into a digital signal.

seafloor spreading The process of forming new ocean floor at oceanic ridges.

secondary energy source Energy in a form that can be distributed easily but is manufactured by using a raw energy resource such as a fossil fuel or wind. Examples of secondary energy sources are electricity, hot water used in heating systems, and steam.

selective absorption   Some materials absorb some forms of electromagnetic radiation but not others. For example, glass absorbs infrared but is transparent to visible light.

sidereal day   The time taken for the Earth to rotate 360°: 23 hours and 56 minutes.

signal   Information carried through a communication system, for example, by an electromagnetic wave with variations in its amplitude or frequency, or being rapidly switched on an off.

slope   The slope of a graph is a measure of its steepness. It is calculated by choosing two points on the graph and calculating: the change in the value of the y-axis variable/the change in the value of the x-axis variable. If the graph is a straight line, you can use any two points on it. If it is curved, you can estimate the slope at any chosen point, using two points on the graph close to this point.

solar day   The time taken for the Earth to rotate so that it fully faces the Sun again: exactly 24 hours.

solar eclipse   When the Moon comes between the Earth and the Sun, and totally or partially blocks the view of the Sun as seen from the Earth's surface.

Solar System   The Sun and objects that orbit around it – planets and their moons, comets, and asteroids.

source   An object that produces radiation.

spectrometer   An instrument that divides a beam of light into a spectrum and enables the relative brightness of each part of the spectrum to be measured.

spectrum   One example is the continuous band of colours, from violet to red, produced by shining white light through a prism. Passing light from a flame test through a prism produces a line spectrum.

speed of light   300 000 kilometres per second – the speed of all electromagnetic waves in a vacuum.

speed of recession   The speed at which a galaxy is moving away from us.

star life cycle   All stars have a beginning and an end. Physical processes in a star change throughout its life, affecting its appearance.

static electricity   Electric charge that is not moving around a circuit but has built up on an object such as a comb or a rubbed balloon.

strong (nuclear) force   A fundamental force of nature that acts inside atomic nuclei.

Sun   The star nearest Earth. Fusion of hydrogen in the Sun releases energy, which makes life on Earth possible.

supernova   A dying star that explodes violently, producing an extremely bright astronomical object for weeks or months.

tectonic plate   Giant slabs of rock (about 12, comprising crust and upper mantle) that make up the Earth's outer layer.

telescope   An instrument that gathers electromagnetic radiation, to form an image or to map data, from astronomical objects such as stars and galaxies. It makes things visible that cannot be seen with the naked eye.

tension   A material is in tension when forces are trying to stretch it or pull it apart.

thermistor   An electric circuit component whose resistance changes markedly with its temperature. It can therefore be used to measure temperature.

transformer   An electrical device consisting of two coils of wire wound on an iron core. An alternating current in one coil causes an ever-changing magnetic field that induces an alternating current in the other. Used to 'step' voltage up or down to the level required.

transmitted (transmit)   When radiation hits an object, it may go through it. It is said to be transmitted through it. We also say that a radio aerial transmits a signal. In this case, transmits means 'emits' or 'sends out'.

transverse wave   A wave in which the particles of the medium vibrate at right angles to the direction in which the wave is travelling. Water waves are an example.

ultraviolet (UV)   Electromagnetic waves with frequencies higher than those of visible light, beyond the violet end of the visible spectrum.

umbra   An area of total darkness in a shadow. For example, places in the Moon's path where the Earth completely blocks off sunlight.

Universe   All things (including the Earth and everything else in space).

unstable   The nucleus in radioactive isotopes is not stable. It is liable to change, emitting one of several types of radiation. If it emits alpha or beta radiation, a new element is formed.

velocity   The speed of an object in a given direction. Unlike speed, which only has a size, velocity also has a direction.

velocity–time graph   A useful way of summarising the motion of an object by showing its velocity at every instant during its journey.

vibration   Moving rapidly and repeatedly back and forth.

volcano   A vent in the Earth's surface that erupts magma, gases, and solids.

voltage   The voltage marked on a battery or power supply is a measure of the 'push' it exerts on charges in an electric circuit. The 'voltage' between two points in a circuit means the 'potential difference' between these points.

voltmeter   An instrument for measuring the potential difference (which is often called the 'voltage') between two points in an electric circuit.

wave speed   The speed at which waves move through a medium.

wavelength   The distance between one wave crest (or wave trough) and the next.

work   Work is done whenever a force makes something move. The amount of work is force multiplied by distance moved in the direction of the force. This is equal to the amount of energy transferred.

X-ray   Electromagnetic wave with high frequency, well above that of visible light.

Index

Appendices

Relationships

You will need to be able to carry out calculations using these mathematical relationships.

P1 The Earth in the Universe

distance travelled by a wave = wave speed × time

wave speed = frequency × wavelength

P3 Sustainable Energy

energy transferred = power × time

power = voltage × current

$$\text{efficiency} = \frac{\text{energy usefully transferred}}{\text{total energy supplied}} \times 100\%$$

P4 Explaining motion

$$\text{speed} = \frac{\text{distance}}{\text{time}}$$

$$\text{acceleration} = \frac{\text{change in velocity}}{\text{time taken}}$$

momentum = mass × velocity

change of momentum = resultant force × time for which it acts

work done by a force = force × distance moved in the direction of the force

amount of energy transferred = work done

change in gravitational potential energy = weight × vertical height difference

kinetic energy = ½ × mass × velocity^2

P5 Electric circuits

power = voltage × current

$$\text{resistance} = \frac{\text{voltage}}{\text{current}}$$

$$\frac{\text{voltage across primary coil}}{\text{voltage across secondary coil}} = \frac{\text{number of turns in primary coil}}{\text{number of turns in secondary coil}}$$

P6 Radioactive materials

Einstein's equation: E = mc^2 where E is the energy produced, m is the mass lost, and c is the speed of light in a vacuum

P7 Studying the Universe

$$\text{power of a lens} = \frac{1}{\text{focal length}}$$

$$\text{magnification of a telescope} = \frac{\text{focal length of objective lens}}{\text{focal length of eyepiece lens}}$$

Hubble equation: speed of recession = Hubble constant × distance

Einstein's equation: E = mc^2 where E is the energy produced, m is the mass lost and c is the speed of light in a vacuum

For a fixed mass of gas:

pressure × volume = constant *at constant temperature*

$$\frac{\text{pressure}}{\text{temperature}} = \text{constant } \textit{for constant volume}$$

$$\frac{\text{volume}}{\text{temperature}} = \text{constant } \textit{at constant pressure}$$

Units that might be used in the Physics course

length: metres (m), kilometres (km), centimetres (cm), millimetres (mm), micrometres (μm), nanometres (nm)

mass: kilograms (kg), grams (g), milligrams (mg)

time: seconds (s), milliseconds (ms)

temperature: degrees Celsius (°C); Kelvin (K)

area: cm^2, m^2

volume: cm^3, dm^3, m^3, litres (l), millilitres (ml)

speed and velocity: m/s, km/s, km/h

acceleration: metres/second^2 (m/s^2)

force: newtons (N)

momentum: kilogram-metres/second (kg m/s)

energy/work: joules (J), kilojoules (kJ), megajoules (MJ), kilowatt-hours (kWh), megawatt-hours (MWh)

electric current: amperes (A), milliamperes (mA)

potential difference/voltage: volts (V)

resistance: ohms (Ω)

power: watts (W), kilowatts (kW), megawatts (MW)

frequency: hertz (Hz), kilohertz (kHz)

information: bytes (B), kilobytes (kB), megabytes (MB)

radiation dose: sieverts (Sv)

distance (astronomy): parsecs (pc)

power of a lens: dioptres (D)

Prefixes for units

nano	micro	milli	kilo	mega	giga	tera
one thousand millionth	one millionth	one thousandth	× thousand	× million	× thousand million	× million million
0.000000001	0.000001	0.001	1000	1000 000	1000 000 000	1000 000 000 000
10^{-9}	10^{-6}	10^{-3}	$\times 10^3$	$\times 10^6$	$\times 10^9$	$\times 10^{12}$

Useful data

P1 The Earth in the Universe
speed of light = 300 000 km/s

P2 Radiation and light
speed of light = 300 000 km/s

electromagnetic spectrum in increasing order of frequency:

radio waves, microwaves, infrared, red visible light violet, ultraviolet, X-rays, gamma rays

P3 Sustainable energy
mains supply voltage: 230 V

P4 Explaining motion
a mass of 1 kg has a weight of 10 N on the surface of the Earth

P5 Electric circuits
mains supply voltage: 230 V

P6 Radioactive materials
speed of light (c) = 300 000 000 m/s

P7 Studying the Universe
solar day = 24 hours

sidereal day = 23 hours 56 minutes

age of the Universe: approximately 14 thousand million years

absolute zero of temperature: 0 K = −273 °C

Electrical symbols

junction of conductors		ammeter	
switch		voltmeter	
primary or secondary cell		motor	
battery of cells	or	generator	
		fixed resistor	
		variable resistor	
power supply		light dependent resistor (LDR)	
lamp		thermistor	

OXFORD
UNIVERSITY PRESS

Great Clarendon Street, Oxford OX2 6DP

Oxford University Press is a department of the University of Oxford.
It furthers the University's objective of excellence in research,
scholarship, and education by publishing worldwide in

Oxford   New York

Auckland   Cape Town   Dar es Salaam   Hong Kong   Karachi
Kuala Lumpur   Madrid   Melbourne   Mexico City   Nairobi
New Delhi   Shanghai   Taipei   Toronto

With offices in
Argentina   Austria   Brazil   Chile   Czech Republic   France   Greece
Guatemala   Hungary   Italy   Japan   Poland   Portugal   Singapore
South Korea   Switzerland   Thailand   Turkey   Ukraine   Vietnam

Oxford is a registered trade mark of Oxford University Press
in the UK and in certain other countries.

British Library Cataloguing in Publication Data.

Data available.

ISBN 978-0-19-913842-5

10 9 8 7 6 5 4 3 2 1

Printed in Spain by Cayfosa-Impresia Ibérica.

Paper used in the production of this book is a natural, recyclable product made
from wood grown in sustainable forests. The manufacturing process conforms to
the environmental regulations of the country of origin.

Acknowledgements
The publisher and authors would like to thank the following for their permission
to reproduce photographs and other copyright material:
P13t: David Taylor/Science Photo Library; **P13b:** John Howard/Science Photo
Library; **P14:** Johan Ramberg/Istockphoto; **P18l:** Frank Zullo/Science Photo
Library; **P18r:** Detlev Van Ravensswaay/Science Photo Library; **P19t:** NASA/CXC/
STScI/JPL-Caltech/Science Photo Library; **P19b:** Mark Garlick/Science Photo
Library; **P20:** Jerry Lodriguss/Science Photo Library; **P21:** Zooid Pictures; **P22:**
Chris Butler/Science Photo Library; **P23t:** Tony Hallas/Science Photo Library;
P23b: NASA/ESA/STScI/R. Williams, Hdf Team/ Science Photo Library; **P25:**
Colin Cuthbert/Science Photo Library; **P26l:** Jack Sullivan/Alamy; **P26m:** Enzo
& Paolo Ragazzini/Corbis; **P26r:** Sinclair Stammers/Science Photo Library; **P28:**
Bettmann/Corbis; **P34:** James Wardell/Rex Features; **P42l:** NASA/ESA/STScI/R.
Williams, Hdf Team/ Science Photo Library; **P42r:** AZPworldwide/Shutterstock;
P44: Gustoimages/Science Photo Library; **P46:** Trevor Worden/Photolibrary;
P48: Terraxplorer/Istockphoto; **P49t:** NASA/Science Photo Library; **P49b:** Solent
News And Photos/Rex Features; **P50t:** David Turnley/Corbis; **P50m:** Silver-
john/Shutterstock; **P50b:** CNRI/Science Photo Library; **P51:** Mike Hill/Alamy;
P52t: Image Source/Alamy; **P52bl:** Astier - Chru Lille/Science Photo Library;
P52br: Mark Sykes/Science Photo Library; **P53:** University of Oxford- Division
of Public Health and Primary Health Care; **P55t:** Janine Wiedel/Janine Wiedel
Photolibrary/Alamy; **P55b:** Ted Kinsman/Science Photo Library; **P59t:** Philip
Lange/Shutterstock; **P59bl:** Martin Muránsky/Shutterstock; **P59bm:** The Flight
Collection/Alamy; **P59br:** Photofusion Picture Library/Alamy; **P60t:** British
Antarctic Survey/Science Photo Library; **P60b:** George Steinmetz/Science Photo
Library; **P61:** Victor De Schwanberg/Science Photo Library; **P64l:** David J. Green -
lifestyle themes/Alamy; **P64r:** lebanmax/Shutterstock; **P66t:** Loskutnikov/
Shutterstock; **P66b:** Jerry Mason/Science Photo Library; **P72:** David J. Green -
lifestyle themes/Alamy; **P74:**Stephen Strathdee/Istockphoto; **P76t:** Christopher
Walker/Shutterstock; **P76b:** ronfromyork/Shutterstock; **P77t:** Tonylady/

Shutterstock; **P77b:** Jeffrey Van Daele/Shutterstock; **P78:** Cecile Degremont/
Look At Sciences/Science Photo Library; **P79:** Martin Moxter/Photolibrary;
P80: Ben smith/Shutterstock; **P81t:** Ilya Akinshin/Shutterstock; **P81b:** Foment/
Shutterstock; **P82tl:** Rex Features; **P82tr:** Harald Tjøstheim/Dreamstime;
P82ml: Olexa/Fotolia; **P82mr:** Frances A. Miller/Shutterstock; **P82b:** Jonathan
Feinstein/Shutterstock; **P84:** Ieva Geneviciene/Shutterstock; **P85:** Sheila Terry/
Science Photo Library; **P87:** Barry Batchelor/PA Photos; **P88t:** Paul Rapson/
Alamy; **P88m:** B. S. Merlin/Alamy; **P88b:** © freelights.co.uk 2010; **P89:** ©
2009 Dragonfly; **P91l:** Sean Gallup/Getty Images News/Getty Images; **P91m:**
Ron Giling/Photolibrary; **P91r:** Mark Sykes/Science Photo Library; **P92:** Steve
Allen/Science Photo Library; **P93:** Ria Novosti/Science Photo Library; **P94t:**
Marcomayer/Shutterstock; **P94b:** Pearl Bucknall/Robert Harding/Rex Features;
P95t: D. Kusters/Shutterstock; **P95ml:** hjschneider/Shutterstock; **P95mr:**
Rhoberazzi/Istockphoto; **P95b:** Pool/Joao Abreu Miranda/AFP Photo; **P96:**
Victor De Schwanberg/Science Photo Library; **P97:** Penimages/Dreamstime;
P99: Martin Bond/Science Photo Library; **P102:** Htjostheim/Dreamstime;
P104: Paul Mckeown/Istockphoto; **P106:**Vakhrushev Pavel/Shutterstock; **P107:**
Photodisc/Photolibrary; **P108l:** NASA; **P108r:** Greg McCracken/Istockphoto;
P109: P B images/Alamy; **P110l:** Fine Shine/Shutterstock; **P110r:** Power And
Syred/Science Photo Library; **P113:** Kenneth Eward/Biografx/Science Photo
Library; **P114:** Santiago Cornejo/Shutterstock; **P121:** Offside/Rex Features;
P122: EuroNCAP Partnership; **P128:** Adrian Houston/Photolibrary; **P132:** P B
images/Alamy; **P134:** Blackred/Istockphoto; **P136:** NOAA Central Library; OAR/
ERL/National Severe Storms Laboratory (NSSL)/NOAA Photo Library; **P142:**
Anthony Redpath/Corbis; **P143:** Zooid Pictures; **P146:** Masterfile; **P147:** Ilya
Zlatyev/Shutterstock; **P153:** Phillppe Hays/Rex Features; **P156:** Sheila Terry/
Science Photo Library; **P158t:** Anthony Vizard/Eye Ubiquitous/Corbis; **P158b:**
T H Longannet/Scottish Power; **P159l:** Hink General 01 HR/British Energy;
P159m: Nicholas Bailey/Rex Features; **P159r:** Sub_station3/Scottish Power;
P162: Ilya Zlatyev/Shutterstock; **P164:** Efda-Jet/Science Photo Library; **P166t:**
Kletr/Shutterstock; **P166bl:** Health Protection Agency/Science Photo Library;
P166br: Radiation Protection Division/Health Protection Agency/Science Photo
Library; **P167t:** Davies and Starr/Stone/Getty Images; **P167bl:** Matthias Kulka/
Corbis; **P167br:** Peter Thorne, Johnson Matthey/Science Photo Library; **P170:**
Amrit G/Fotolia; **P173t:** Geoff Tompkinson/Science Photo Library; **P173b:**
Krzysztof Slusarczyk/Shutterstock; **P175:** Julia Hedgecoe; **P176t:** Josh Sher/
Science Photo Library; **P176b:** Health Protection Agency/Science Photo Library;
P177: Health Protection Agency/Science Photo Library; **P182:** Prof. J. Leveille/
Science Photo Library; **P185t:** Jerry Mason/Science Photo Library; **P185b:**
Ria Novosti/Science Photo Library; **P186t:** Steve Allen/Science Photo Library;
P186b: Keith Beardmore/The Point/British Nuclear Fuels Limited; **P189:** ITER/
Science Photo Library; **P192:** Geoff Tompkinson/Science Photo Library; **P196t:**
Adam Hart-Davis/Science Photo Library; **P196b:** Detlev Van Ravensswaay/
Science Photo Library; **P197bl:** Jerry Lodriguss/Science Photo Library; **P197br:**
AFP; **P198t:** Dr Juerg Alean/Science Photo Library; **P199t:** Herman Heyn/
Science Photo Library; **P199b:** The British Library Board; **P202t:** NASA; **P202b:**
Royal Astronomical Society/Science Photo Library; **P203:** Victor Habbick
Visions/Science Photo Library; **P206t:** Dr Fred Espenak/Science Photo Library;
P206b: George Bernard/Science Photo Library; **P207b:** Dr Fred Espenak/Science
Photo Library; **P208l:** Royal Astronomical Society/Science Photo Library; **P208r:**
D.A. Calvert, Royal Greenwich Observatory/Science Photo Library; **P209m:**
NASA; **P209bl:** Eckhard Slawik/Science Photo Library; **P209br:** X-ray: NASA/
UIUC/Y.Chu et al., Optical: NASA/HST; **P211tl:** Detlev Van Ravensswaay/Science
Photo Library; **P211tr:** Nasa/Science Photo Library; **P211bl:** Image courtesy of
NRAO/AUI; **P211br:** NASA/JPL-Caltech/Science Photo Library; **P212:** Arthur S.
Aubry/Riser/Getty Images; **P213:** Justin Quinnell/NASA; **P218r:** Science Photo
Library; **P220:** David Parker/Science Photo Library; **P223l:** Dr Juerg Alean/
Science Photo Library; **P223r:** Konstantin Remizov/Shutterstock; **P225:** Dr Seth
Shostak/Science Photo Library; **P226t:** P227t: NRAO/AUI/NSF/ Science Photo
Library; **P227bt:** MRAO; **P228:** NASA; **P230:** ESO/G.Hüdepoh; **P231tl:** ESO/Jose
Francisco Salgado; **P231tr:** ESO/H.H.Heyer; **P231b:** ESO/H.H.Heyer; **P234t:**
NASA; **P234b:** NASA; **P235tl:** NASA/ESA/STScI/AURA/A. Evans (University Of
Virginia, Charlottesville; NRAO; Stony Brook University)/Science Photo Library;
P235tr: NASA/ESA/Hubble Heritage Team/STScI/Science Photo Library; **P237:**
European Space Agency/Science Photo Library; **P239:** NASA/ESA/H. Richer,
UBC/STSCI/Science Photo Library; **P241:** Harvard College Observatory/Science
Photo Library; **P244t:** Richard J. Wainscoat, Peter Arnold Inc./Science Photo
Library; **P244m:** JAXA/NASA; **P244bl:** NASA/ESA/STScI/A. Fruchter, Ero Team/
Science Photo Library; **P244bm:** Nasa/Esa/Jpl/Arizona State University; **P244br:**
Nasa/Esa/Stsci/J.Bahcall, Princeton Ias/Science Photo Library; **P245:** Noao/
Science Photo Library; **P246:** Emilio Segre Visual Archives/American Institute
Of Physics/Science Photo Library; **P248:** NASA/Science Photo Library; **P249m:**
Physics Dept., Imperial College/Science Photo Library; **P249b:** European Space
Agency/Science Photo Library; **P250t:** Royal Observatory, Edinburgh/Aao/
Science Photo Library; **P250b:** Dr Juerg Alean/Science Photo Library; **P256:**
Segre Collection/American Institute Of Physics/Science Photo Library; **P260:**
Robert Gendler/NASA; **P261:** SOHO/ESA/NASA/Science Photo Library; **P264:**
European Southern Observatory/Science Photo Library; **P268:** European Space
Agency/Science Photo Library; **P270:** David A. Hardy, Futures: 50 Years In
Space/Science Photo Library; **P272:** NASA/Science Photo Library; **P275b:** David
Parker/Science Photo Library. **P281:** NASA.

Illustrations by IFA Design, Plymouth, UK, Clive Goodyer, and Q2A Media.

The publisher and authors are grateful for permission to reprint the following copyright material:

Although we have made every effort to trace and contact all copyright holders before publication this has not been possible in all cases. If notified, the publisher will rectify any errors or omissions at the earliest opportunity.

Project Team acknowledgements
These resources have been developed to support teachers and students undertaking the OCR suite of specifications GCSE Science Twenty First Century Science. They have been developed from the 2006 edition of the resources.
We would like to thank David Curnow and Alistair Moore and the examining team at OCR, who produced the specifications for the Twenty First Century Science course.

Authors and editors of the first edition
We thank the authors and editors of the first edition, David Brodie, Peter Campbell, Simon Carson, John Lazonby, Robin Millar, Stephen Pople, David Sang, and Carol Tear.
Many people from schools, colleges, universities, industry, and the professions contributed to the production of the first edition of these resources. We also acknowledge the invaluable contribution of the teachers and students in the pilot centres.
The first edition of Twenty First Century Science was developed with support from the Nuffield Foundation, The Salters Institute, and the Wellcome Trust. A full list of contributors can be found in the Teacher and Technician Resources.

The continued development of Twenty First Century Science is made possible by generous support from
- The Nuffield Foundation
- The Salters' Institute